ADDRESSES

OF THE

NEWLY-APPOINTED PROFESSORS

OF

COLUMBIA COLLEGE,

WITH AN INTRODUCTORY ADDRESS

BY WILLIAM BETTS, LL.D.

FEBRUARY, 1858.

NEW YORK:
BY AUTHORITY OF THE TRUSTEES.
1858.

WYNKOOP, HALLENBECK & THOMAS, PRINTERS,
No. 113 Fulton street, New York.

RESOLUTIONS OF THE TRUSTEES.

AT a meeting of the Trustees of Columbia College, held on the 22d of June, 1857:

Resolved, That it be referred to a committee of three, of which the President shall be Chairman, to make such arrangements as they may deem expedient in reference to the inauguration of the newly-appointed Professors at the opening of the next College Course.

The President, the Rev. Dr. Spring, and the Rev. Dr. Haight were appointed the Committee.

Mr. Jones and Mr. Zabriskie were subsequently added to the Committee.

On the 17th September, 1857, the Committee made a Report, and it was

Resolved, That the Chairman of the Committee on the Course be requested to prepare the address of the Trustees, mentioned therein.

At a meeting of the Trustees of the College, held on the 1st of March, 1858, it was

Resolved, That the Chairman of the Committee on the Course, and the several Professors, be requested to furnish copies of the addresses lately delivered by them, and that the same be printed and published under the direction of the Committee.

CONTENTS.

INTRODUCTORY ADDRESS

AT THE

INAUGURATION

OF THE

Newly-Appointed Professors

OF

COLUMBIA COLLEGE,

BY WILLIAM BETTS, LL. D.,

February, 1858.

INTRODUCTORY ADDRESS.

Before the several professors, who have lately assumed chairs of instruction in Columbia College, shall address themselves to you, the Trustees have thought it fitting, on their own behalf, to offer a few preliminary remarks. The principal object of our meeting at this time is, the introduction to the public of those learned gentlemen. On ordinary occasions, this introduction is made by means of an Inaugural Discourse, delivered by the incumbent, on commencing the functions of his office; and it seldom happens that more than a single individual, at any one time, is called upon to perform the duty. We have now, however, to celebrate no common inauguration. Several individuals are about to lay before you their views and the objects of their teaching in the various departments committed to them; and this, under circumstances of deep interest, and peculiar importance. The rapid progress of events, for a few past years, has operated powerfully upon the College, and wrought a mighty change in its prospects and designs. Those events have pushed from its foundations the old fabric of the College, and swept away every trace of its material existence. This revolution was attended with adequate compensation. An accession to its means came with its change of position; and that brought with it an accession of responsibility. This is not just the time to enlarge on the character of that responsibility, or of the mode in which it affected the minds of the Trustees. It is enough, at this moment, to say, that they felt it, and they acted upon it; they wish that the Public should know that they did feel it, and that they did act upon it; and, with this view, they have directed, in connection with the present celebration, to use their own words, "An Address to be prepared from the Trustees, setting forth, clearly and fully, the history of the recent changes in and enlargement of the course of studies; and their purposes and hopes in regard to their future operations." The Trustees having thought fit to select one of their own number for the performance of that duty; and having committed it to him who now speaks to you, he will proceed to discharge it, with as much brevity as is consistent with its faithful performance.

The prominent object before the Public, undoubtedly, is the removal of the College, and its position in its present locality. The considerations and motives which preceded that removal, the hopes which attend, and the completed designs which may follow it, are not so apparent to them; and, in making a sketch of the late changes, it may not be unfitting to refer briefly to the recent removal of the Institution from its former grounds.

Undoubtedly this removal is attended with painful recollections, as well as pleasing anticipations. At the time the old building was erected, one hundred years ago, it stood in the midst of pleasant fields, on the banks of the lordly Hudson. Tranquillity and silence were around it. The groves of the Academy were removed from the strife of trade, and the tumult of the forum; the life of contemplation was not brought breast to breast with the life of action, but each pursued, apart, its own appropriate course: and the students, while they daily, in their silken gowns and tasselled caps, proceeded to the isolated halls of their venerated mother, felt themselves a distinct and peculiar class. The conviction rested upon them, although it might find no definite embodiment either in idea or expression, that their quiet and modest occupation was one of high import; that their daily intercourse with the spirits of the great men, whose intellect had illuminated the world, was fitting them for high intercourse thereafter; and that they were becoming qualified to be the leaders and benefactors of mankind.

The progress of trade soon filled those fields with habitations, and removed from the shore of the College lawn the ripples of the gentle river. The Academic caps disappeared from the brows of the youthful wearers; and houses, and ware-rooms, and streets crowded around, and stretched onwards, and pressed upon the College precincts, until the voice of learning was almost stifled by the clamor of business; its atmosphere mingled with the atmosphere of trade; its objects began to be regarded as of secondary import; and the halls of study lost that calm repose, without which study can never be profitably prosecuted.

Although, for some years before the removal took place, it became evident that it must occur at no distant period, its actual removal was yet made sooner than it was expected. The opening of a new street, directly in front of the building, taking from it all retirement and privacy, and the current of trade, which poured steadily just in that direction, forced it to retreat. The old recollections which clustered around it, of early friendships, and joyous sports, and youthful aspirations, were banished with regret; the old associations connected with the halls, and the Lecture rooms, and the honored faculty, and even the neglected discipline, with pain were broken. It was painful, but it was proper; and the Government of the College did not hesitate to do it. They had not waited for the last moment, but had already taken such measures as were fitting for the occasion; and when the event took place, they were, so far as it was in their power to be so, prepared for it. It was anticipated that the same circumstances which compelled the removal of the College would eventually enlarge its income. It could not be known when, or to what extent, this enlargement would take place; nor what amount of expenditures might become needful; but it was plain that, after all expenditures, there would yet be an accession to its means. The College felt that it was incumbent on it to improve, if possible, the quality of its usefulness, and certainly its quantity; and it took timely measures to provide for every possible contingency.

The property of the College, it is generally known, is composed of the noble donation, by Trinity Church, of the tract on which the fabric lately stood; and of another tract of land, the gift of the State, formerly known as the Botanic Garden, near the position of the present College.

The first grant was made "to aid in founding, erecting and establishing a college, and promoting the education of youth in the liberal arts and sciences;'

and was upon the condition that the President should be in communion with the English Church, and that certain prayers should be used in the daily service of the chapel. This grant forms the principal part of the active property of the College. Large sums have lately been received from the sale of a portion of the property. A considerable amount of the avails of those sales is required for adequately establishing the institution in the upper part of the city, and for putting the State grant in a condition to produce a revenue. The accommodations now in use for instruction are intended to be temporary. The outlay made in preparing them for use has much exceeded what was intended; a heavy expenditure has been incurred in regulating the Botanic Garden, from which an essential part of the future revenue of the College is expected to be derived; this expenditure has absorbed large sums received from the sales of the real estate; and great prudence and economy are necessary to secure the advantages of education, at which the Trustees most anxiously are aiming. These observations, respecting the means of the College, are not precisely in the chronological order in which they should be in relation to the changes made; but, inasmuch as they form the basis upon which the other measures are to be constructed, they are here introduced, that the connection of its action, on those measures, may not be broken. You will observe, therefore, that when the Trustees of the College began to move in this matter, they knew that very large expenses must be incurred both in the construction of the buildings and in the regulation of the State grounds; they knew that the expenses of regulating the land would eventually be paid by the land itself; but they did not know when they might expect revenue from it; they did not know how much the portion of the church property to be sold would produce; they knew that, upon the whole, there would be an increase of revenue, but they could not anticipate its amount, or the time at which it would take place; and therefore it became material that any plan of enlarged operations should admit of expansion in proportion to the means to be acquired.

Actuated by these considerations, the Trustees, on the third day of October, 1853, appointed a committee, who were instructed "to inquire whether it was expedient to take any and what measures, for the removal of the College; and, in the event of such removal, whether any and what changes ought to be made in the under-graduate course; and whether it would be expedient to establish a system of university education, in addition to such under-graduate course, either in continuation thereof or otherwise; and that they should report fully as to the principles and details of any plan that they might recommend;

and whether, in their opinion, it could be successfully carried into execution; and in connection therewith, that they consider whether, for the more effectual carrying out such plan, and extending the benefits of the institution, it ought to afford rooms and commons, or rooms alone for resident students, or ought to have its seat isolated."

These instructions, it may be observed, covered the whole subject of the higher branches of learning; and, by directing the principles of any plan to be detailed, it necessarily opened an inquiry into the objects and ends of a thorough education, as well as the best means of attaining those objects and ends. The Committee entered upon their duties with promptness; and at the next meeting of the Board of Trustees, in November, 1853, they made a preliminary report, in which they stated, as the groundwork of their future proceedings, their conviction, that the proper business of a College education was the cultivation of the human intellect in all its parts and functions, with a view to a full development of the mental and moral qualities: generally, to form and give direction to the mind, without reference to any specific future employment.

The Committee knew that this view did not command popular sympathy. They knew that the public generally, unaccustomed to look upon the mind except in connection with the body, and to regard it as a machine for promoting the pleasures, the conveniences, or the comforts of the latter, might be dissatisfied with a system of education in which they were unable to perceive the direct connection between the knowledge imparted and the advantages to be gained. They hoped that some means might be devised for satisfying, in some measure, this demand; but, in seeking this object, they were admonished, by experience, authority and reason, not to diminish, in the slightest degree, the high value which was placed on the right acquisition of the Greek and Latin Classics. With respect to the establishment of a post-graduate university system in addition to the under-graduate course, they were not prepared to say more, than that they regarded it favorably in those respects in which it might be practicable: but that the design was not free from serious difficulties; that the subject had occupied the minds of learned men in connection with the English Universities, but hitherto without effect; that the Medical and Theological Schools had done much, perhaps all that could at present be done in that direction; but in regard to higher jurisprudence, and the sciences and their applications, much might possibly be done by the College.

The Committee likewise recommended the imme

diate removal of the College; but, although exertions were made to attain that object, they were fortunately, as subsequent events proved, unsuccessful.

The general principle of Collegiate Education having been thus briefly, but decidedly exhibited to the Board of Trustees, and no dissenting opinion having been expressed, it became proper to invoke the advice of the several members of the Faculty, whose acquirements and experience peculiarly qualified them to afford aid at this juncture; and without whose harmonious co-operation with the Trustees no success could be expected in the proposed operations. Most of the Faculty gave written responses to the inquiries addressed to them; and it was satisfactory and gratifying that their concurrence with the views of the Committee, as to the fundamental principles and true ends of education, was entirely unanimous.

The answers of the Faculty to the inquiries made had not, however, been immediate; and, in the meanwhile, a full report was made to the Board of Trustees, by the same Committee, on the 24th of July, 1854, comprehending all the subjects which had been referred to them, and reporting fully the principles which had guided them in the adoption of the plan recommended. That plan, as eventually modified, will soon be explained. It was contained in a syllabus or outline of a statute; and, as the revenues of the College were not then in a condition to authorize an immediate expansion, an opportunity was allowed for examination and criticism, and for receiving from the Faculty the mature results of their reflections. The answers of several of the Faculty having been at length received, and the Board of Trustees having had a full opportunity for deliberation, the Committee, on the 4th of June, 1855, again brought the subject before the Trustees, and professed themselves ready to report a statute, at any time that the Board might desire, and be prepared to receive it.

On the 12th of January, 1857, the time had arrived, when from the necessity of removal, and the probability of an augmented revenue, the Board of Trustees were prepared for action; and accordingly, on that day, they directed the Committee on the course, "to bring in the full statute to comprehend the whole scheme of College and University instruction contemplated by their former report."

The requisition was promptly obeyed, and the full statute prepared and reported on the second day of March. In the brief interval, the site on which the old College stood had been sold, and the removal of the institution was to take place in May.

For two or three years preceding this period, arrangements had been going on for the erection of proper and permanent buildings on the Botanic Garden grounds; but the uncertainty attending the eventual arrangement of those grounds, and other circumstances which could not be controlled, suspended the prosecution of the design. A variety of proposals had from time to time been made for removal to temporary buildings; but the transfer of a large institution, with its library, apparatus, and necessary paraphernalia, was no easy matter. This subject might have created a serious embarrassment, had not the offer of the buildings, formerly occupied by the Deaf and Dumb Institution, provided a mode of accommodation, which promised to be both efficacious and economical.

The Trustees now set about effecting their purpose in earnest. By the month of July the statute was modified, altered and completed, and assumed its present shape; and, about the commencement of the Collegiate term in autumn, four professors and one associate had been added to the body of instructors.

You have been told that, in directing the preparation of the statute, the Committee were instructed to comprehend in it the whole scheme of Collegiate and University studies contemplated by their former report. The phrase "University studies" was one of mere convenience; perhaps not very accurate, and was intended to denote that instruction which might be imparted after graduation.

The studies denominated Collegiate are well understood. They comprise the various branches known as the Classical, Philosophical, Historical, Belles-lettres, Mathematical, and in some degree Scientific. The latter term, however inappropriate as an exclusive name, has been assumed by that peculiar branch of human learning which comprehends the nature, operation and laws of Matter. That branch has been, and is, perpetually expanding by new discoveries. If it were expected that it should be included, much beyond the elements, in the usual collegiate course, then a useful college education would be simply impossible. There had been indications abroad, that, notwithstanding its evident impracticability, this was expected. It would fatigue you to enter into any detail of the suggestions or discussions on this subject. It may suffice, for this occasion, to say that an effort was made to satisfy, as far as possible, all demands, and that the plan now adopted, and about to go into operation, was upon the whole, after full consideration, regarded as the best which could be fallen upon, for an experiment.

That plan adopts in substance the former collegiate curriculum to the close of the Third or Junior year, with adaptations to the future studies, both sub-graduate and post-graduate. At the commencement of the Fourth or Senior year, the studies assume a wider scope, and comprehend a variety of topics. These studies, too numerous to be pursued in one or two courses, even in the most elementary manner, are distributed into three departments, in order that they may be prosecuted with some hope of advantage. Up to this point of college life, the end in view is mainly to discipline and invigorate the mind, and to enlighten and purify the heart. Now, the object is to apply this intellectual light and vigor to the permanent acquisition of knowledge; to emancipate the student gradually from the trammels of catechetical teaching, and to prepare him for the higher and more arduous efforts of self-instruction. With this view, three departments are constructed, which are termed Schools of Letters, of Science, and of Jurisprudence; the first of which has reference to general improvement; the two latter to specific objects, as indicated by their names. On entering the Senior year, each student may select either of these schools. Should he neglect to make a selection, he continues in the Classical or School of Letters.

After graduation, the same schools are proposed to be continued for two years. A reference to the proposed course of instruction will show that they comprehend a large circle of human learning. The instruction in these schools is not to be confined to the graduates of the college. It is open to the whole world. A sufficient body of teachers is provided to commence the undertaking. A nucleus is presented for a great university, adapted and prepared to meet all the wants of the community. If there be really that demand for the acquisition of knowledge which has been supposed, it may here be satisfied. If there be in fact no such demand, or such only to a limited extent, time will soon develop the truth. It is indeed hoped that the graduates of the college, animated by a noble and inspiring love of learning, will not fail to take advantage of the proposed means of instruction thus afforded to them, and that others will gradually be drawn to join them.

The progress of the undertaking may be slow; it may be unsuccessful. The slowness of its progress need, however, not to produce despair. Most things that are valuable and lasting are slow in progression. Time and experience will, however, soon demonstrate the utility of the attempt; and it is so devised, that it may be expanded, contracted, or discontinued without difficulty.

Among the late changes, it may be observed, that the Chapel exercises have been modified, and greater solemnity imparted to the service; the Library has been stimulated; large accessions have been made to the Chemical and Physical Departments; the Astronomical Department forwarded, and measures taken for the establishment of an Observatory; and, in general, liberal contributions have been made to the requirements of all the chairs. The price of tuition has been reduced nearly one-half. The division of the classes into sections gives an opportunity for more thorough instruction. Moreover, a Commission of Inquiry has been instituted, for gathering information from every accessible quarter of this country, having in view the general advancement of learning, and enforcement of discipline, the results of which have not yet been made known to the Trustees.

The revenue of the College is yet limited; its enlargement is prospective, of course uncertain, and under any circumstances much below the common supposition. Should the effort now in progress be successful, the Trustees propose to add, from time to time, every necessary appliance for the advancement of learning, to the extent of the means enjoyed by them; but they feel it a duty not to lavish those means, and, by heedlessly exhausting them, defeat the hopes of the true friends of education.

A brief history has thus been offered of the late changes in the course of study, of their enlargement, and of the purposes and hopes of the Trustees in regard to their future operations. Before dismissing this subject, you will pardon a few general remarks having reference to the principles comprehended in the recent measures of the College.

The young student, when he presents himself for admission into College, is just emerging from boyhood; and, before he completes the portion of time to be passed within its walls, will be, essentially, a man. The class of youth who are sent here are, for the most part, a select class; selected from their supposed promise for the future; selected by the fond expectations of loving parents; or to qualify themselves rightly to occupy the position in society for which Providence appears to have designed them. They are to occupy the higher positions of society: those young men are thereafter to form a leaven, which is to spread its influence throughout the community, either for evil or for good. If their tendencies be well directed, they will become a blessing to themselves, to their immediate companions, and to the world in which they may move. Should their tendencies be ill-directed, they will be equally a curse. With the vivacity and buoyancy of youth, with its joyous, exuberant and not easily restrained overflow of spirits, they generally bring with them an apprehension of truth, an instinct of honor, and an appreciation of justice, which, if discreetly managed, will lead and keep them in the right direction. If there be exceptions; if there be found some ignoble spirits who have strayed among them, it does not require much observation to detect them, and they should at once be rooted out. This young band, who annually present themselves, asking to be carried through the most critical portions of their lives, and who confidingly throw themselves into the protection of the College; who come, in a measure, divested of the unsleeping and anxious carefulness which has watched them from the cradle up to this period, when they seek to be instructed to walk alone; to go forth from the College walls armed with the panoply of virtue and of learning; to meet the masses of evil which they will be sure to encounter in later life; have a right to all the thought, all the intelligence, and all the experience that can be brought to bear upon their situation. It is true that the period between their entrance into College life and their departure from it is short; but it is the very heart of their life; it is just the period which gives color to their future, and stamps it for good or for evil in this world, and it may be in the world to come. The responsibility which rests upon giving a right direction during this period, is just in proportion to the greatness of the results; and no right-thinking man, whether among the Trustees or the Faculty, can fail to feel the graveness of the charge which is laid upon him. Whatever other duties may devolve upon the authorities of the College, this one is clear, that they are bound to exercise their best energies and their best judgment for the benefit of the youth entrusted to their care; and, although it may be incumbent on them to extend, as far as practicable, the circle of human learning, and to bring as many as possible within its operation, the other is the most pressing and paramount obligation.

Perhaps it may be said, that the means to attain the desired ends are obvious—that they are simply Discipline and Education—words easy to be spoken; difficult to be apprehended; most difficult to be rightly employed. With regard to discipline, can you not see at once the difference which may result from regarding the young as noble, confiding, and honest, or as selfish, narrow, and insincere? In the applications of rewards, can you not perceive the distinction between those which appeal to lofty sentiments, and those which address themselves to sordid feelings? In punishments, between those which are gently but unavoidably applied as a warning to youthful infirmity, and those which are rudely inflicted as a punishment of vice? And yet these are but broad distinctions which separate many varieties. On the Trustees of the College it devolves to prescribe judicious rules; they should look well that those rules be founded on right principles;

but on the Faculty lies a far greater weight of duty. Their daily intercourse with the students is not susceptible of regulation; and it is from their conduct, and their example, that habits of order, diligence, obedience and truth, must be acquired: and as it is written, that spark kindleth spark, and fire answereth to fire, so the young men in their turn should never fail to give to generous confidence a generous response.

With regard to Education, likewise, there are great diversities of opinion, both as respects its object and its means. By some it is regarded as a mere preparation for an establishment in some calling or profession; by others as the guide to the young for the discharge of all their duties in after-life; for thoroughly understanding the nature of every relation that may be thrown upon them, and of applying the highest principles and the greatest power in every position in which they may be placed; never allowing them for a moment to forget that they are heirs of an unending life, and stewards of a priceless trust. The former is the popular idea of collegiate and all other education—the latter is that which influences us here. To educate the intellect, to purify and direct the heart, to train the youthful aspirants to correct motives and designs, to provide them with the means of successfully pursuing any career which they may hereafter select, these are the ends which this College has essentially in view, in its system of sub-graduate instruction. Nor does it adopt this system unaided by the lights of usage, experience, and authority. The gymnasia of Continental Europe and the great universities of the English Islands conduct their instruction upon this principle; and, especially in the case of the latter, with wonderful success. The great thinkers of all times have advocated the cultivation of the mind, as the object of first importance; and the acquisition and application of knowledge as altogether secondary. Of the great thinkers, it will be sufficient to point to one in the ancient, and another in modern times.

In one of his dialogues, Plato represents Socrates, when pointing to the magnificent works that extended and secured the commerce of Athens, enlarged her revenues, and filled her with material comforts, and splendors, and luxuries, as calling all those possessions the merest trifles in comparison with the fundamental virtues; nay, more, as probable causes of future misfortune, and their projectors as authors, not of benefaction, but of calamity.

One of the most brilliant intellects of our own days has not thought this subject beneath his mighty mind. A liberal education is defined by the late Sir William Hamilton to be "an education in which the individual is cultivated, not as an instrument towards some ulterior end, but as an end unto himself alone: in other words as an education, in which his absolute perfection as a man, and not merely his relative dexterity as a professional man, is the scope imme diately in view." To add to these authorities would be easy, but is unnecessary.

Looking at this subject from any point, we may safely conclude that the instruction in the College, covering the period of life between boyhood and manhood, and forming the bridge by which we pass from home into the world, is one of peculiar importance. To the College is committed the mind of the future man at this critical time; and it is the proper duty of the College to direct and superintend the mental and moral culture, and to form the mind or man. Moral and intellectual discipline is the object of Collegiate education. The mere acquisition of learning, however valuable and desirable in itself, is subordinate to this great work. Not only is this the peculiar business of the College, but in the College alone, as a general rule, can this work be performed. The design of a College is, to make perfect the human intellect in all its parts and functions, by means of a thorough training of the intellectual faculties to their full development, and, by the proper guidance of the moral functions, to a right direction. To form the mind, is, in short, the high design of education as sought in a College course.

But this College, in the enlargement of its course and of its objects, does not propose to stop short here. Thus far the intellectual faculties have been developed and strengthened; and a right direction given to the moral functions. The youth is now supposed to be competent to begin any task to which his strong inclination, or peculiar disposition, may direct him; and which he may undertake with greater prospect of usefulness than if any peculiar class of powers had been cultivated, to the neglect of others.

To this time have been mainly used the common means of education, which are found in the classics, the mathematics, and the teachings of moral and mental science, in connection with the whole history of man, his thoughts, his relations, duties, deeds and productions; and these, it is thought, properly applied and industriously appropriated, will produce the fairest result of a finished intellectual discipline, and present an accomplished intellect, prepared for any career, competent to encounter any difficulties, in learning, in morals, or in action, and capable, with courage and perseverance, of overcoming or removing them.

Then it is that the post-graduate or university establishment offers its aid in the prosecution of special pursuits, excluding, for the present, the faculties of Theology and Medicine. Then the young student, following the road which has been partially entered in the last year of his college life, may direct his exertions to the particular calling selected for his future career; and then it is that Science, with its speculations, and discoveries, and applications, may be profitably studied.

Permit a single observation more, and you shall no longer be detained from the discourse which is to follow.

It may be remarked, that while augmentations have been made to all the other chairs of the college, those of the Greek and Latin alone are left as they were. Do not attribute this to any diminution of esteem for those venerable and noble languages. When the full course shall be in operation, adequate aid will not be wanting to those chairs; but their incumbents are both able and willing to undertake all the present labors, heavy though they be. This College ever has acted upon the principle, that the very best means of intellectual training may be found in the learned languages. For this purpose, these languages are successfully employed from the period when the child first acquires an easy mastery of his mother tongue for ordinary purposes, down to the time when the intellect becomes vigorous in early manhood. During this period, the learned languages, by their novelty, regular structure and musical beauty, awaken a love of study, command the attention, strengthen the memory, improve the reasoning faculty and the judgment; call into action exactness, comparison, invention, self-reliance, and all the other faculties, besides laying up a rich store of beautiful images, noble sentiments, worthy examples, and a mass of facts which, dwelling in the midst of harmonious and perfect tongues, purify the heart, exalt and ennoble the principles, create and cultivate a refined taste, enlarge the understanding, arouse a love of freedom and of virtue; and, in short, fill the whole man with a power of appreciating excellence.

Wretched, indeed, would be the day for this institution, should she lose her proud position as the Classical College of the country; saddened the hopes of her sons, should she become indifferent to the precious treasures of those ancient people. Nowhere, since the creation of this earth, when first the Almighty Spirit spread forth his wings above the mingled chaos, and with the fiat of his word called it to Light, to Order, and to Life, nowhere and at no time has the world beheld so marvellous an exhibition of intellectual power, as in the Grecian people, during the short period of Grecian domination; nowhere so exquisite an appreciation of beauty in all its forms, material and mental, and so wonderful a power of producing it; nowhere such models of intelligence in every branch of human acquisition and human inquiry, which never have been equalled in after-times, and which serve to show of what the human mind is capable.

Nor were the Romans less wonderful in their peculiar character. Deriving many of their acquisitions from the Greeks, and second to the Greeks alone, they far transcend all other people, and throw all other histories wholly in the shade. To count the gifts we have received from these two nations, whether in their laws, their literature, their customs, their arts, or the precious legacy of the Sacred Scriptures, transmitted in the Grecian tongue, we could not tell where to begin, or where to end. So abundant, indeed, are their productions, so precious their treasures, that the same eulogy, which would elsewhere be extravagance, becomes to them but truth. Never, therefore, may this venerable institution become insensible to the value of classic learning; never may she cease from its copious fountains to draw exuberant supplies; never, never, may she forget that, saving the gift of the Sacred Writings, in these old treasures of Greece and Rome are garnered the most precious stores of deep philosophy, unequalled wisdom, of unrivalled eloquence, of poetic excellence; and there, too, those marvels of artistic beauty, elevating the imagination, refining the sentiments and purifying the heart, which age after age admires and wonders at, and which are endued with a grace and loveliness beyond the rivalry and almost the imitation of modern times. Never may our College cease to be a seminary in which such things are taught, and through which a knowledge of them may in some measure be attained: and then, when, in her after-course, she is preparing her sons for the busy tumults of life, or unfolding to them the strange secrets of the material world, explaining their operations, and applying their powers to the good of man, or walking with them in the high regions of the heavenly lights, she may say to them, My sons, I have tried in all things to perform my duty; I have opened to you the treasures of the past and of the present; I have sought to impart to you not only learning but understanding; I have taught you to regard knowledge no otherwise than as "a rich storehouse for the glory of God and the relief of man's estate;" and I charge you, as you value the privileges of the past and the aspirations of the future, I charge you, never to apply it to any lower purposes.

CHEMISTRY;

AN

INAUGURAL ADDRESS,

Delivered before the Trustees

OF

COLUMBIA COLLEGE,

February 4, A. D. 1858,

BY CHARLES A. JOY, PH. D.,

PROFESSOR OF CHEMISTRY IN COLUMBIA COLLEGE.

ADDRESS.

In assuming the duties of the Chair to which the Trustees of Columbia College have done me the honor to appoint me, it is proper that I should sketch the rise and progress of the science which I am called upon to teach, and vindicate its claims to be regarded as among the most important branches of human knowledge.

Chemistry was not accorded a place as a distinct science, until within the memory of men still living. The name, it is true, had existed for centuries. With its prefix *Al*, it carries us away back to its learned Arabian professors, and suggests to us the discoveries they made and the uses they made of them. But it was not until the world had profited by Black's researches into Fixed Air, and Priestly had made his immortal discovery of Oxygen; not until Cavendish had decomposed the ancient element of Water, and Lavoisier, while separating the pure jewel of scientific truth from the rubbish of ages, and extending, by his own investigations, the boundaries of Chemical knowledge, had given order and system to the results of the labors of his predecessors and contemporaries that Chemistry was entitled to an independent position among the sciences.

It is pretty evident that the Egyptians, at a very early period, had far outstripped their neighbors in the pursuit of knowledge, and that, too, of various kinds, but more especially in the various departments of what we now call Natural Philosophy. Hence, it is common to attribute to them considerable acquaintance with Chemical facts. But whatever may have been the extent of this knowledge, it would seem that it was confined to the priests. In this they were by no means alone, for such was then, and for very long after, the case among all nations. The priesthood was the great repository of learning of every sort; the religion, the laws, and the government were all more or less in their hands, as well as the knowledge of the facts and principles on which were based their systems of both mental and physical philosophy, the greater part of which they seem to have carefully concealed from the popular mind.

It was to these Egyptians that the Greeks were indebted for much of their knowledge. Then, as now, the true seekers after knowledge and truth left their homes to look in other lands for additions to their stores. And it was when thus engaged that Pythagoras, and Solon, and Herodotus, and Plato, and others of the mighty and noble spirits of Greece gained the information and acquired the experience with which they returned to their own land, to reproduce, and vastly improve upon, what they had learned from the sages of Egypt. From the Greeks, again, the Romans received their most valuable lessons, in all learning, and in this, our particular branch of knowledge, as in all others. Nor were they at all negligent of their acquisitions, for they soon applied their energy and newly acquired skill in such a way as to make very considerable contributions to Chemical Science.

True, they both had some strange notions about the use to be made of their Chemical knowledge, which, however, were thoroughly utilitarian. The notion of transmuting the base metals into gold was such an one. And this is supposed to have existed among the Greeks. One interpretation of the Golden Fleece is, that it is a mythical expression for a parchment on which had been written a description of the process of making gold. Aristotle recognized four elements, viz.: Earth, Air, Fire and Water, though he also taught the existence of still another substance of a more ethereal nature, which he called the fifth essence, or, as it is expressed in Latin, *essentia quinta* (quintessence); and this fifth element played an important part in the controversies of after-ages.

These strange notions, as we call them, did not vanish with Grecian eloquence, or fall with Roman power. It was in following out such notions that the Alchemists of the Middle Ages kept on searching for the Philosopher's Stone, that great transmuter of all base things into veritable gold. Though they never found the stone, we have this day to thank them for the many valuable discoveries which they did make, and of which the Chemists of the 18th and 19th centuries were able, to much purpose, to avail themselves.

By these Alchemists Sulphuric Acid was discovered more than a thousand years ago; and to them we also owe Muriatic Acid, Nitric Acid, Ammonia, the Fixed Alkalies, Alcohol, Ether, and many Alloys of the Metals. They accounted for everything in what we would call a supernatural way; they looked upon bodies not merely as inorganic masses, but they taught the presence of a spirit in every combination, and, in accordance with their belief and their teaching, they gave names which still remain in daily use among us. To this we owe such names as Spirit of Wine, Spirit of Salt, Spirit of Ether, and the like. At a later period, Van Helmont, following somewhat in their path, gave to all aeriform bodies the name of Gheist or Spirit, and from which we derive the modern word Gas.

As they taught the existence of a spirit in bodies so they taught that these bodies were affected in various ways. The baser metals they spoke of as diseased; Brass was diseased Gold; Quicksilver, diseased Silver; and so they accounted for all the phenomena of nature in a manner which, at the present day, it is difficult to look upon as having ever been regarded as philosophical. For example, they said that the cause of the falling of a body is its weight, and that weight is the tendency of a body to fall. A stone falls, said they, because it is heavy, that is, because it has a tendency to downward motion. Opium, they said, produces sleep, because it is a body to which belongs a sleep-producing property. The caustic properties of Potash were said to be due to a *something* which they called *Causticum*. And in this manner they were prepared to give an explanation of every phenomenon of nature.

For fourteen hundred years no Alchemist ventured to dispute the views of his predecessors. An unqualified submission to the traditions of the past characterized this long period. But the founding of the Universities, the discovery, first, of the passage round the Cape of Good Hope, and then of the American Continent, gave an impulse to everything, and occasioned a greater interchange of knowledge among the nations, whilst the seizure of Constantinople by the

Turks, in A. D. 1453, scattered a knowledge of the Arts and Sciences throughout Western Europe. The invention of Printing and the events of the Reformation put an end to the blind obedience to the authorities of the past, which had so long prevailed.

Paracelsus was one of the first to impart a new direction to Chemical researches. He affirmed that the decayed forces of the human constitution might be indefinitely extended by means within the reach of man. His dreams are matters of amusement now, but the value of his influence, in breaking up the old alchemistic theories, cannot be overrated. He was a Harbinger of our great Science, and a Pioneer in its work of discovery, and therefore entitled to more than a passing notice in its History.

Philip Aureolus Theophrastus Paracelsus Bombastus von Hohenheim, as he styled himself, was born in Einsiedeln, Switzerland, A. D. 1493. His father was a Physician, and early instructed him in Medicine, Astrology, and Alchemy. Paracelsus was a great traveler, visiting nearly every part of Europe, and finally retiring to Saltzburg, where he died, A. D. 1541. Though he was not old when he died, it was generally believed that he had discovered the Elixir Vitæ, and his tomb became an object of superstitious veneration. Even to the present day the stones about it are worn away by the numbers who come to pray for friends afflicted with disease. I once went to visit the laboratory of this remarkable man. A tablet on the wall indicated the house, and some flowers in a window showed that it was inhabited. I readily obtained permission to examine the room in which Paracelsus had compounded those strange mixtures which are not dreamt of even in the quackery of our day. Scarcely a remnant of the old hearth and flues remained; but it was interesting to stand upon the spot on which the first great opposition was made to the Alchemistic theories, and from which Chemistry first started upon its path of usefulness to mankind. For two hundred years, the attention of Chemists having been withdrawn, through the influence of Paracelsus, from the search after the Philosopher's Stone, investigators took the direction of Pharmaceutical Chemistry, and some important discoveries were made.

Stahl, who died A. D. 1685, was the first to promulge the Phlogistic theory, which occupied such an important place in the studies of a hundred years.

Phlogiston, according to him, was present in every Chemical phenomenon. He taught, for instance, that Phosphorus, when burnt, loses its Phlogiston, and that the white acid, which is the result of the combustion, if mixed with charcoal and distilled, yields Phosphorus, because the coal gives back the lost Phlogiston to the Phosphorus. And it is remarkable that, although the substance, in burning, is increased in weight, they still clung to the idea that it lost Phlogiston, though this Phlogiston, they maintained, was possessed of *levity*, and thus made a body lighter. This theory was stoutly maintained for a hundred years, until overturned by the grandest of all Chemical discoveries, that of Oxygen, which was made by Joseph Priestley, on the first day of August, A. D. 1774.

A dispute in the French Academy, between Cadet and Baumé, about the red-oxide of Mercury, led to Priestley's making some researches into the properties of this compound. He concentrated the solar-rays upon the red-precipitate, and preserved the gas which was evolved; he applied a lighted taper to this gas, and from that moment the discovery had been made. According to his own account, it was accident which led to the discovery, but accident only accords such favors to those who deserve them. The man who had discovered nine gases, who had invented all the apparatus necessary to prepare and study them, could well have laid claim to this immortal discovery.

Chemistry, as a distinct science, dates from this discovery. It is nearly of the same age as our Republic, and, in its way, it has made equal progress to greatness. The immortal discoverer of Oxygen sought refuge in this country, and died at Northumberland,

Pennsylvania, on the sixth day of February, A. D. 1804, in the 71st year of his age.

"Westward the Star of Empire takes its way."

May this prove true in Science and Art, as well as in political advancement.

There is something sublime in the thought of being the discoverer of such an element as Oxygen. When we contemplate its abundance, its necessity to the very existence of all animated nature, and the part it plays everywhere, we are struck with amazement that it should have remained so long unknown, and are able to appreciate more fully the importance of its discovery.

Independently of Priestley, and about the same time, the famous Swedish Chemist, Scheele, made the same discovery; but to Priestley is due the honor of having laid the foundation upon which the whole superstructure of Chemistry has since been built.

We must not pass from the last century into our own, without mentioning, with due honor, the name of Lavoisier, for the world is chiefly indebted to the genius of this great man for the invention of that new Nomenclature in Chemistry which marked a brilliant era in the History of the Science.

This remarkable man, "possessed of fortune sufficient to secure to him all the gratifications of luxury, all the splendors of a princely establishment, gave his time and his enthusiasm to the Science of Chemistry. But, unfortunately, he took such an active part in the first scenes of the French Revolution as to render himself obnoxious to Robespierre. He was pursued to his retreat, where he was carrying on a train of magnificent Chemical Experiments, and hurried thence to the scaffold." There were few victims of those bloody times who could not have been better spared than this illustrious man. Strange fate of the two great founders of Chemistry, the one beheaded, the other driven an exile to a foreign land.

It was at the beginning of the present century that Berzelius, the most illustrious of all the Chemists that ever lived, first appeared as an independent investigator in the field of Physical Science. Volta had given to the world the Battery which bears his name, and the whole Scientific community was in excitement upon the subject. Berzelius entered upon investigations by aid of the Battery, and he was the first to discover the thread which has led to the explanation of many of the mysteries in Chemical combinations. The account of the investigations, in the course of which he systematized bodies in reference to their positive and negative poles, was published in *Gehlen's Journal*, during the year 1803. Three years later (in 1806), Sir Humphrey Davy published an account of similar researches, in which he made no mention of Berzelius, and for which he received from Napoleon the prize of three thousand francs, for the most valuable researches in Voltaism.

After the discovery of the Metals of the Alkalies, in the year 1807, Berzelius continued his researches, and was the first to use Mercury at one of the poles of the Battery. It was not, however, with Voltaism alone that Berzelius, in his earlier years, occupied his time. The great Mineralogist, Hisinger, induced him to turn his attention to the quantitative analysis of minerals. He acknowledged to Rose, in after-years, that these investigations were undertaken more to oblige Hisinger than for any particular interest they had for himself. But, after the discovery of the great law of Chemical proportions, the case was different.

Berzelius was compelled to earn his living as a practicing physician. This, at times, gave a particular turn to his investigations. Hence, we find him examining Medicinal Springs; or establishing, at Stockholm, the manufacture of artificial Mineral Waters; or, as a physician, early giving his attention to Physiological and Organic Chemistry, and it is from this circumstance that we are indebted to him for many new and beautiful processes of Analysis. Then again we find him, in some of his earliest efforts, busy in the examination of Silica and Cast Iron.

During the first decennary of this century, Berzelius was induced, by the general interest in Galvanism, by the influeuce of his friend Hisinger, and by his own necessities as a Physician, to make scientific researches; but, after the idea of Chemical proportions was started, he devoted all his energies to Chemistry; he soon put forth the Law of these proportions, and upon that Law he founded all the subsequent experiments and researches of his life. He examined with the greatest care, and by different modes of Analysis, a vast number of Chemical compounds, and was thus led to discover many methods of analysis, which are still pursued. These researches and their results were published in the year 1810. At the time they were carried on, Re-agents were scarcely to be had in Sweden, and Berzelius was compelled to make them for himself; even Alcohol and the most ordinary Acids were prepared in his own Laboratory. The extraordinary spectacle was, at this time, presented to the world, of a Philosopher at work in his kitchen, making researches which were destined to revolutionize Chemical Science, and for which the world could not have adequately compensated him if it had erected for him a Laboratory of solid gold; while by his side, at the same hearth, his faithful servant, Anne, was preparing his frugal meal. He introduced more accurate Balances; the use of smaller amounts of substances for analysis; the lamp which bears his name; platinum-crucibles; Swedish filtering-paper; funnels; beaker glasses; and many pieces of apparatus which now seem to us very common and simple indeed. He was also the first to make the Laboratory a light and cheerful study instead of a dark and dismal cellar. As it was necessary for him to economize in everything, he took lessons in Glass-blowing, and learned the trade of the Joiner, so that he could make nearly all of his apparatus himself. It was in the year 1815 that Berzelius introduced the symbols which are now employed in the place of the Alchemistic figures, which are retained only to designate the planets, and was thus enabled to express the Chemical composition of different bodies by Formulæ. Dalton had undertaken, in the year 1808, to establish some simple method for expressing the composition of bodies, but it was not so practical as the one prepared by Berzelius, and is, at present, scarcely known. It was not until he had occupied ten years in examining the elements and their combinations that Berzelius was able to publish, in the year 1818, his Tables, containing the Atomic-weights of nearly two thousand simple and compound bodies.

It has already been said that Berzelius, at an early period, gave some attention to Organic Chemistry. The first important analyses in this department were made by Thénard and Gay-Lussac, in the year 1811.

But, in the year 1814, Berzelius published a paper on this subject, in which he applied the Law of Chemical-proportions to Organic Bodies. He found that Organic Acids, and even indifferent substances, formed compounds of fixed proportions with Organic Oxides. This originated the Radical theory, and through its application we are able to ascertain the Atomic-weights of Organic substances.

Since the death of Berzelius many Radicals, proposed by him as hypothetical, have been confirmed by actual discovery.

The discovery and investigation of the properties of Selenium was one of the greatest works of Berzelius. Excepting the discovery of Selenic acid, in the year 1827, by his pupil, Mitscherlich, very little has since been added to our knowledge of this metal. "The investigations upon this element," says Rose, "were made with half an ounce of substance, a part of which was lost by the carelessness of a servant." The beauty of the work can only be compared with that of the investigations upon Iodine by Gay-Lussac. Berzelius examined one hundred and twenty Salts of Sulphur, many of them quantitatively. The publication of his famous Hand-book occasioned the examination of a vast number of substances, the composition of which had never been satisfactorily determined. It is rarely within the ability of one man to leave such a monument of greatness behind him. As long as Chemistry endures, this book will claim a place in every Laboratory, and the name of Berzelius will be mentioned with honor.

I cannot now speak of his many other contributions to Chemical Science. His last great work was the Examination of Meteorites, in which, without success, he sought to discover some new elements. His age and frequent headaches did not admit of his working in the Laboratory. He complained of his eyes and of loss of memory. As he could not carry on his practical labors, he devoted so much the more time to the theory and literature of the Science. After he became permanent Secretary of the Royal Academy he was successful in introducing yearly reports of the progress of the Sciences. He undertook, as his part, Physics, Mineralogy, Geology, and all branches of Chemistry.

His first Report, which was made thirty-six years ago, was contained in a very thin octavo volume. At the present time such a volume could not contain the index of the discoveries made in one year, nor would it, now, be within the power of one mind to gather in and store up the vast harvest which the wide field covered by these Sciences at this day yields.

When Berzelius visited Germany, in the year 1845, he was everywhere received by the students with addresses, processions, and other tokens of honor. His pupils, already great and renowned, flocked round him, and he had no occasion to be ashamed of any of them. One day when, in company with Humboldt, Mitscherlich, the two Roses, Wœhler, Ehrenberg, and von Buch, he drove out to the environs of Berlin, they stopped to examine a Boulder outside the gate. What a group was standing around this erratic mass! each contributing, of his vast knowledge, to resolve some question connected with its history. Berzelius could give its exact Chemical composition; Mitscherlich apply his Law of Isomorphism; Wœhler tell whether any of the Elements discovered by him were contained in it; Henry Rose prescribe the best methods of analysis; Gustavus Rose measure accurately every crystal; Leopold von Buch explain its Geological origin; Ehrenberg find former life in its minutest grains; while the comprehensive genius of Humboldt could sum up the case for all sides, and pronounce a decision to which every one would cheerfully submit. A piece of this Boulder was broken off, and carried to Gœttingen by Wœhler; it afterwards found its way to this country, and is now preserved at Amherst College.

I have dwelt thus long upon the name and works of Berzelius, because, in speaking of Chemistry, that name and those works are so interwoven with the history of that Science, that neither can be fairly presented without displaying the other. His influence, too, is not limited to his personal labors, but still lives in the pupils he trained. It is remarkable how many eminent men received their first impulse in his Laboratory. Mitscherlich, Henry and Gustavus Rose, Magnus, Gmelin, Wœhler and Turner were among the number.

It has been my good fortune to sit at the feet of some of these pupils of Berzelius, and through them I have endeavored to draw inspirations from his instructions; and I feel that I am but paying a small portion of the debt due to a great teacher, by whose lessons I have thus profited, as well as performing an appropriate duty on this public inauguration of my appointment to this Chair of Chemistry, in thus rendering my humble homage to his exalted genius.

Time will not allow me to dwell any longer on the History of Chemistry. The sketch already given shows its very recent origin, and its rapid growth. As an Abstract Science, its progress has been wonderful. It has almost approached the Mathematics in its definite precision, and it may be said to be daily assuming a more and more strictly Mathematical form. But it is not confined to abstract principles; it has descended into the daily walks of practical life. Some of its greatest discoveries have become so familiar in their practical applications, that we almost forget their scientific origin. In the trite allusions to the triumphs of the Steam-engine and of the Electric Telegraph, the obligations we owe to Chemistry are well nigh forgotten. How few are there who, beholding the beautiful light which sheds its lustre around us this evening, remember that to the labors of men of Science we owe this great social blessing. "When the earth brought forth grass, the herb yielding seed, and the fruit tree yielding fruit, and GOD had said 'Let there be light,' and there was light," each ray of sunshine, as it impinged upon tree or flower, was caught up, and held in close embrace, and, with the decay of the plant, was carried down into the earth, changed by slow degrees in form, and there kept a close prisoner, until the genius of man could deliver it from its bondage. These imprisoned sunbeams, after the lapse of ages, are set at liberty at the beck of man, and here we see them gushing forth, from many a jet, turning our night into brilliant day. If Chemistry had contributed no other blessing to the world, this alone would have entitled it to the gratitude of man.

The practical utility of Abstract Science is not recognized by the bulk of mankind. They become so familiar with the Material results, that they forget altogether its Scientific origin, and sometimes look with scorn upon studies which are now laying the foundation for practical good for a future generation. As Baron von Liebig well remarks, "In our schools mere children are now taught truths, the attainment of which has cost immense labor and indescribable effort. They smile when we tell them that an Italian Philosopher wrote an elaborate Treatise to prove that the snow found upon Mount Ætna consists of the same substance as the snow upon the Alps of Switzerland, and that he heaped proof upon proof that both these snows, when melted, yielded water possessed of the same properties.

"When a school-boy takes a glass full of liquid, and, placing a loose piece of paper over it, inverts the glass without spilling a drop of its contents, he only astonishes another child by his performance, and yet this is the identical experiment which renders the name of Torricelli immortal. Our children have more correct notions of Nature and Natural phenomena than had Plato! They may laugh at the errors committed by Pliny in his Natural History." Indeed, no Science exhibits more beautifully the harmony between Abstract truth and Practical utility; and there is none in which thorough cultivation is more directly beneficial to the world at large. The Chemists of the world are accumulating a great store of knowledge, the utility of which, to the human race, can hardly be overrated. Its far-reaching results extend to almost every article of human use; and Scientific truths, which now seem without any practical utility, will, without doubt, yield rich fruits to another generation.

But, of all countries where the cultivation of Science would produce the most useful results, our own stands conspicuous. With the natural wealth so richly spread over our wide Empire, in all that the bounteous earth produces, and the hidden stores she carries in her bosom, our countrymen need but the key which Science gives to enable them to unlock their treasure-house. But their impatience for results, their excessively practical character, make them miss the success they might securely attain by pursuing the proper method with patience and perseverance. How many wild schemes of speculation might have been avoided, how many fortunes saved from ruin by a proper application of Scientific knowledge! No greater benefaction could be bestowed on our country than to diffuse everywhere within its borders sound Scientific principles, and any measures tending to this end must contribute directly and largely, to the public good. Our country needs not only the widest diffusion, but also the highest grade of Science. How can it be attained? By devising and putting into operation the means adequate to produce the desired result.

The great want of this country is a University where Science can be taught far beyond the usual College course, where the students may be led into the profounder regions of the interpretation of phenomena, as well as into the practical application of Science to the daily wants of man.

We present the extraordinary spectacle of a nation possessing unbounded wealth, and yet affording no aid to that very Science to which we are chiefly indebted for onr material success. A University is the great educational want of America. In other lands the government extends its protection and aid not only to Elementary Schools and Colleges, but also to Universities. Some part of what is done by governments abroad, is effected by private munificence here, at home. All honor to men like Lawrence, Peabody, Astor, Lennox, Nott, Delavan, and to one in our own city, who has erected a massive structure to be devoted to Science and Art, and which, notwithstanding another corporate name, will ever be known as The Cooper Institute.

We need more such men to aid us in carrying out the work; but the founding and building up of a Unversity is beyond the power of individual effort. May we not consider it Providential that Columbia College is placed in a position which will enable her to confer this great blessing on our land? Placed at the great centre of all the interests of our country, with the heaving vitality which is around her, and with ample means, we may safely indulge the hope that the great scheme, which her Trustees have formed, will be carried out to completion, and that, at some not distant day, a great University will be established, which will afford a home for those scholars who are now driven to foreign lands to perfect themselves in Science. But great as the present advantages and the prospective wealth of Columbia may be, the aid of every lover of education will be needed to secure success. I feel confident that this aid will not be wanting, that the dishonor of America will soon be wiped away, and that we shall see a University worthy of the name, and worthy of our country. When that day comes our young men will find other professions besides the few which they can now pursue, and they will discover, in Science, charms far more attractive than can be found in the frivolous amusements and ignoble dissipations by which they are, now, too often led away. Chemistry, alone, is comprehensive enough to receive all who may wish to approach her. The separate fields of Pharmacy, Agriculture, Analysis, Physiology, Technology, and Organic Chemistry, have been but partially explored. We are upon the threshold of discovery in them all, and the progress of Science stands still for want of laborers.

On surveying this vast and ever-spreading field, ever opening, as it is, into new regions with each new accession of knowledge, I am awe-struck, and feel almost rebuked for my presumption in undertaking, single-handed, to introduce the pupils of Columbia College into this boundless domain of Science. I look forward, however, to the better day of the University, when I shall have the co-operation and support of fellow-laborers in this great field, each cultivating his own portion of the domain, each adding to the great common stock of Scientific Truth, and all raising still higher the renown of this venerable Institution, and rendering it more and more a glory and blessing to our country.

INAUGURAL ADDRESS

OF

FRANCIS LIEBER, LL. D.,

Professor of History and Political Science.

Delivered on the 17th of February, 1858.

ADDRESS.

The author, requested by the Board of Trustees to prepare a copy of his inaugural address for publication, has given the substance, and in many places his words, as originally delivered, so far as he remembered them; but, some of his friends in the Board, having advised him not to restrict himself in the written address, to the limits necessary for one that is spoken, he has availed himself of this liberty, in writing on topics so various and comprehensive, as those that legitimately belong to the branches assigned to him in this institution. The extent of this paper will sufficiently indicate this.

GENTLEMEN OF THE BOARD OF TRUSTEES:

We are again assembled to do honor to the cause of knowledge—to that sacred cause of learning, inquiry and rearing to learn and to inquire; of truth, culture, wisdom, of humanity. Whenever men are met together to reverence a great cause or to do homage to noble names, it is a solemn hour, and you have assigned a part in this solemnity to me. I stand here at your behest. No one of you expects that I should laud the sciences which form my particular pursuit, above all others. Every earnest scholar, every faithful student of any branch, is a catholic lover of all knowledge. I would rather endeavor, had I sufficient skill, to raise before you a triumphal arch in honor of the sciences which you have confided to my teaching, with some bas-reliefs and some entablatures, commemorating victories achieved by them in the field of common progress; taking heed however that I do not fall into the error of attempting to prove "to the Spartans that Hercules was a strong man."

Before I proceed to do the honorable duty of this evening, I ask your leave to express on this, the first opportunity which has offered itself, my acknowledgment for the suffrages which have placed me in the chair I now occupy. You have established a professorship of political science in the most populous and most active city in the widest commonwealth of an intensely political character; and this chair you have unanimously given to me. I thank you for your confidence.

Sincere, however, as these acknowledgments are, warmer thanks are due to you, and not only my own, but I believe I am not trespassing when I venture to offer them in the name of this assemblage, for the enlargement of our studies. You have engrafted a higher and a wider course of studies on your ancient institution which in due time may expand into a real, a national university, a university of large foundation and of highest scope, as your means may increase and the public may support your endeavors. So be it.

We stand in need of a national university, the highest apparatus of the highest modern civilization. We stand in need of it, not only that we may appear clad with equal dignity among the sister nations of our race, but on many grounds peculiar to ourselves. A national university in our land seems to have become one of those topics on which the public mind comes almost instinctively to a conclusion, and whose reality is not unfrequently preceded by prophetic rumor. They are whispered about; their want is felt by all; it is openly pronounced by many until wisdom and firmness gather the means and resolutely provide for the general necessity. There is at present an active movement of university reform prevailing in most countries of Europe; others have institutions of such completeness as was never known before, and we, one of the four leading nations, ought not to be without our own, a university, not national, because established by our national government; that could not well be, and if it were, surely would not be well; but I mean national in its spirit, in its work and effect, in its liberal appointments and its comprehensive basis. I speak fervently; I hope, I speak knowingly; I speak as a scholar, as an American citizen; as a man of the nineteenth century in which the stream of knowledge and of education courses deep and wide. I have perhaps a special right to urge this subject, for I am a native of that city which is graced with the amplest and the highest university existing. I know, not only what that great institution does, but also what it has effected in times of anxious need. When Prussia was humbled, crippled, and impoverished beyond the conception of those that have never seen with their bodily eyes universal destitution and national ruin, there were men left that did not despair, like the foundation walls of a burnt house. They resolved to prepare even in those evil days, even in presence of the victorious hosts, which spread over the land like an inundation in which the ramified system of police drew the narrow-meshed seine for large and small victims—even then to prepare for a time of resuscitation. The army, the taxes, the relation of the peasant to the landholder, the city government and the communal government—all branches of the administration were reformed, and, as a measure of the highest statesmanship, the moral and intellectual elevation of the whole nation was decided upon. Those men that reformed every branch of government resolutely invigorated the mind of the entire realm by thorough education, by an all-pervading common school system, which carries the spelling-book and the multiplication table to every hut, by high schools of a learned and of a polytechnical character, and by universities of the loftiest aim. The universities, still remaining in the reduced kingdom were reformed, and a national university was planned, to concentrate the intellectual rays and to send back the intensified light over the land. It was then that men like Stein, one of the greatest statesmen Europe has produced, and the scholar-statesman William Humboldt—his brother Alexander went to our Andes—and Niebuhr, the bank officer and historian, and Schleiermacher, the theologian and translator of Plato, and Wolf, the enlarger of philology and editor of Homer, with Buttman the grammarian, and Savigny, the greatest civilian of the age, and Fichte and Steffens the philosophers, these and many more less known to you, but not less active, established the national university in the largest city of Prussia for the avowed purpose of quickening and raising German nationality. All historians as well as all observing cotemporaries are agreed that she performed her part well. In less than seven years that maimed kingdom rose and became on a sudden one of the leading powers in the greatest military struggle on record, calling for unheard of national efforts, and that great system of education, which rests like a high and long arch on the two buttresses, the common school and the university, served well and proved efficient in the hour of the highest national need; and, let me add, at that period when the matrons carried even their wedding rings to the mint, to exchange them for iron ones with the inscription: Gold I gave for Iron, the halls of that noble university stood mute. Students, professors, all, had gone to the rescue of their country, and Napoleon honored them by calling them in his proclamations, with assumed contempt, the school-boy soldiers. They fought, as privates and as officers, with the intelligence and pluck of veterans and the dash of patriotic youth, and when they had fought or toiled as soldiers toil, in the day, many of them sang in the nightly bivouac those songs, that swell the breasts of the Germans to this hour.

We are, indeed, not prostrated like Prussia after the French conquest, but we stand no less in need of a broad national institution of learning and teaching. Our government is a federal union. We loyally adhere to it and turn our faces from centralization, however brilliant, for a time, the lustre of its focus may appear, however imposingly centred power, that saps self-government, may hide for a day the inherent weakness of military concentrated polities. But truths are truths. It is a truth that modern civilization stands in need of entire countries; and it is a truth that every government, as indeed every institution whatever is, by its nature, exposed to the danger of gradually increased and, at last, excessive action of its vital principle. One-sidedness is a universal effect of man's state of sin. Confederacies are exposed to the danger of sejunction as unitary governments are exposed to absorbing central power—centrifugal power in the one case, centripetal power in the other. That illustrious predecessor of ours, from whom we borrowed our very name, the United States of the Netherlands ailed long with the paralyzing poison of sejunction in her limbs, and was brought to an early grave by it, after having added to the stock of humanity the worshipful names of William of Orange, and de Witt, Grotius, de Ruyter and William the Third.* There is no German

* Every historian knows that William of Orange, the founder of the Netherlands' republic, had much at heart to induce the cities of the new union to admit representatives of the *country;* but the "sovereign" cities would allow no representatives to the farmers and landowners, unless noblemen, who, nevertheless, were taking their full share in the longest and most sanguinary struggle for independence and liberty; but the following detail, probably, is not known to many. The estates of Holland and West Friesland were displeased with the public prayers for the Prince of Orange, which some high-Calvanistic ministers were gradually introducing, in the latter half of the seventeenth century, and in 1663, a decree was issued ordaining to pray first of all "for their noble high mightinesses, the estates of Holland and West Friesland, as the true sovereign, and only sovereign power after God, in this province; next, for the estates of the other provinces, their allies, and for all the deputies in the assembly of the States General, and of the Council of State."

"Separatismus," as German historians have called the tendency of the German princes to make themselves as independent of the empire as possible, until their treason against the country reached "sovereignty," has made the political history of Germany resemble the river Rhine, whose glorious water runs out in a number of shallow and muddy streamlets, having lost its imperial identity long before reaching the broad ocean.

among you that does not sadly remember that his country, too, furnishes us with bitter commentaries on this truth; and we are not exempt from the dangers common to mortals. Yet as was indicated just now, the patria of us, moderns, ought to consist in a wide land covered by a nation, and not in a city or a little colony. Mankind have outgrown the ancient city-state. *Countries* are the orchards and the broad acres where modern civilization gathers her grain and nutritious fruits. The narrow garden-beds of antiquity suffice for our widened humanity, no more than the short existence of ancient states. Moderns stand in need of nations and of national longevity, for their literatures and law, their industry, liberty, and patriotism; we want countries to work and write and glow for, to live and to die for. The sphere of humanity has steadily widened, and nations alone can now-a-days acquire the membership of that great commonwealth of our race which extends over Europe and America. Has it ever been sufficiently impressed on our minds how slender the threads are that unite us in a mere political system of states, if we are not tied together by the far stronger cords of those feelings which arise from the consciousness of having a country to cling to and to pray for, and unimpeded land and water roads to move on?

Should we, then, not avail ourselves of so well

proved a cultural means of fostering and promoting a generous nationality, as a comprehensive university is known to be? Shall we never have this noble pledge of our nationality? All Athens, the choicest city-state of antiquity, may well be said to have been one great university, where masters daily met with masters, and shall we not have even one for our whole empire, which does not extend from bay to bay like little Attica, but from sea to sea, and is destined one day to link ancient Europe to still older Asia, and thus to help completing the zone of civilization around the globe? All that has been said of countries, and nations and a national university would retain its full force even if the threatened cleaving of this broad land should come upon us. But let me not enter on that topic of lowering political reality however near to every citizen's heart, when I am bidden by you to discourse on political philosophy, and it is meet for me not to leave the sphere of inaugural generalities.

LADIES AND GENTLEMEN;

This is the first time I am honored with addressing a New York audience, and even if I could wholly dismiss from my mind the words of the Greek, so impressive in their simplicity: It is difficult to speak to those with whom we have not lived—even then I could not address you without some misgiving. The topics on which I must discourse, may not be attractive to some of you, and they cover so extensive a ground, that I fear my speech may resemble the enumeration of the mile-stones that mark the way, rather than the description of a piece of road through cultivated plains or over haughty alps. I, therefore, beg for your indulgence, in all the candor in which this favor can be asked for at your hands.

It is an error, as common in this country as it is great, that every branch of knowledge, if recognized as important or useful, is for that reason considered a necessary or desirable portion of the college course of studies. It is a serious error, but I do not believe that it was committed by the Trustees when they established my chair.

College education ought to be substantial and liberal. All instruction given in a generous college ought to aim at storing, strengthening, refining and awakening the head and heart. It ought to have for its object either direct information and positive transmission of knowledge, for the purpose of applying it in the walks of practical life, or in the later pursuits of truth; or it ought to give the beginnings of knowledge and with them to infuse the longing to enter and traverse the fields which open before the student from the hill-top to which the teacher has led him; or it ought to convey to him the method and skill of study—the scholar's art to which the ancient *Vita brevis ars longa* applies as emphatically as to any other art; or its tendency ought to be the general cultivation and embellishment of the mind, and the formation of a strong and sterling character, Truth and Truthfulness being the inscription on the mansion of all these endeavors.

It is readily understood that all teaching must be within the intellectual reach of the instructed, but it is a grave mistake to suppose that nothing should be placed before the pupil's mind, but what he can actually comprehend in all its details. Life does not instruct us in this manner; the bible does not teach us thus. There is a suggestive instruction, which though occasional, is nevertheless indispensable. It consists in thoughts and topics of an evocative character, giving a foretaste and imparting hope. The power of stimulation is not restricted, for weal or woe, to definition. Suggestive and anticipating thoughts, wisely allowed to fall on the learner's mind, are like freighted sayings of the poet, instinctively recognized as pregnant words, although at the moment we cannot grasp their entire meaning. They fill us with affectionate suspicion. Napoleon was a master of the rhetoric of the camp, as Mackintosh calls it speaking of Elizabeth at Tilbury. His proclamations to the army are described to have had an electrifying effect on every soul in the camp, from the calculating engineer to the smallest drummer boy; yet it is observed that every one of these proclamations, intended for immediate and direct effect, contains portions that cannot have been understood by his hosts. Are we then to suppose that these were idle effusions, escaped from his proud heart rather than dictated for a conscious purpose? He that held his army in his hand as the ancient Cæsars hold Victoria in their palm, always knew distinctly what he was about when his soldiers occupied his mind, and those portions which transcended the common intellect of the camp had, nevertheless, the inspiriting effect of foreshadowed glory, which the cold commander wanted to produce for the next day's struggle. The same laws operate in all spheres, according to different standards, and it is thus that quickening instruction ought not to be deprived of foretokening rays.

Those branches which I teach are important, it seems, in all these respects and for every one, whatever his pursuits in practical life may be. To me have been assigned the sciences which treat of man in his social relations, of humanity in all its phases in society. Society, as I use the term here, does not mean a certain number of living individuals bound together by the bonds of common laws, interests, sympathies and organization, but it means these and the successive generations with which they are interlinked, which have belonged to the same society and whose traditions the living have received. Society is a continuity. Society is like a river. It is easy to say where the Rhine is, but can you say what it is at any given moment? While you pronounce the word Mississippi, volumes of its waters have rolled into the everlasting sea, and new volumes have rushed into it from the northernmost lake Itaska, and all its vying tributaries. Yet it remains the Mississippi. While you pronounce the word America, some of your fellow-beings breathe their last, and new ones are born into your society. It remains your society. How else could I, in justice, be called upon to obey laws, made by lawgivers before I was born and who therefore could not, by any theory or construction, represent me individually? I was not, and therefore had neither rights nor obligations. But my society existed and it exists still, and those are, until repealed, the laws of my society. Society is not arbitrarily made up by men, but man is born into society; and that science which treats of men in their social relations in the past, and of that which has successively affected their society, for weal or woe, is history. Schloezer, one of the first who gave currency to the word Statistik, of which we have formed Statistics, with a somewhat narrower meaning, has well said, History is continuous Statistik; Statistik, History arrested at any given period.

The variety of interests and facts and deeds which history deals with, and the dignity which surrounds this science, for it is the dignity of humanity itself in all its aspirations and its sufferings, give to this branch of knowledge a peculiarly cultivating and enlarging character for the mind of the young.

He that made man decreed him to be a social being, that should depend upon society for the development of his purest feelings, highest thoughts and even of his very individuality, as well as for his advancement, safety and sustenance; and for this purpose He did not only ordain, as an elementary principle, that the dependence of the young of man, and they alone of all mammals, on the protection of the parents, should outlast by many years the period of lactation; and endowed him with a love and instinct of association; and did not only make the principle of mutual dependence an all-pervading one, acting with greater intensity as men advance; but He also implanted in the breast of every human being a yearning to know what has happened to those that have passed before him, and to let those that will come after him know what has befallen him and what he may have achieved—the love of chronicling and reading chronicles. Man instinctively shows the continuity of society long before the philosopher enounces it. The very savage honors the old men that can tell of their fathers and of their fathers' fathers, and tries his hand at record in the cairn that is to tell a story to his children's children. Why do the lonely Icelanders pass their uninterrupted night of whole months in copying Norman chronicles?

As societies rise the desire to know the past as a continuous whole becomes more distinct and the uses of this knowledge become clearer; the desire becomes careful inquiry and collection; mere Asiatic reception of what is given changes into Greek criticism; the love to inform future generations becomes a skill to represent, until history, with the zeal of research, the penetration of analysis, the art and comprehension of representing, and the probity of truth, is seen as the stateliest of all the muses.

So soon as man leaves the immediate interests of the day and contemplates the past, or plans for future generations and feels a common affection with them, he rises to an ennobling elevation. There is no more nutritious pabulum to rear strong characters upon than History, and all men of action have loved it. The great Chatham habitually repaired to Plutarch in his spare half-hours—he had not many—and with his own hands he prescribed Thucydides as one of the best books for his son to read and re-read in his early youth. The biographer of Pitt tells us that while at Cambridge he was in the habit of copying long passages from Thucydides the better to impress them on his mind, as Demosthenes before him had copied the whole. Thucydides is nourishing food. When we read one of our best historical books, when we allow a Motley to lead us through the struggle of the Netherlands, do we not feel in a frame of mind similar to that which the traveler remembers when he left the noisy streets of Rome, with the creaking wine-carts and the screaming street traffic, and enters the Vatican, where the silent, long array of lasting master-works awaits him? Even the contemplation of crime on the stage of history has its dignity as its contemplation on the stage of Shakespeare has. The real science and art of history is the child of periods of action. No puny time has produced great historians. Historians grow in virile periods, and if a Tacitus wrote under the corrupt empire it was Rome in her manhood that yet lived in him and made him the strong historian we honor in that great name. His very despondency is great and he wrote his history by the light which yet lingered behind the setting of Roman grandeur.

There are reasons which make the study of history

peculiarly important in our own day and in our own country. Not only is our age graced with a rare array of historians in Europe and in our hemisphere—I need hardly mention Niebuhr, Ranke, and Neander, and Guizot, and Sismondi, Hallam, Macaulay, and the noble Grote, and Prescott and Bancroft—but, as it always happens when a science is pursued with renewed vigor and sharpened interest, schools have sprung up which in their one-sided eagerness have fallen into serious errors. There was a time when the greatest sagacity of the historian was believed to consist in deriving events of historic magnitude from insignificant causes or accidents, and when the lovers of progress believed that mankind must forget the past and begin entirely new. These errors produced in turn their opposites. The so-called historical school sprung up, which seems to believe that nothing can be right but what has been, and that all that has been is therefore right, sacrificing right and justice, freedom, truth and wisdom at the shrine of Precedent and at the altar of Fact. They forget that in truth theirs is the most revolutionary theory while they consider themselves the conservatives; for what is new to-day will be fact to-morrow and, according to them, will thus have established its historical right.

Another school has come into existence, spread at this time more widely than the other, and considering itself the philosophical school by way of excellence. I mean those historians who seek the highest work of history in finding out a predetermined type of social development in each state and nation and in every race, reducing men to instinctive and involuntary beings and society to nothing higher than a bee-hive. They confound nature and her unchangeable types and unalterable periodicity, with the progress and development as well as relapses of associated free agents. In their eyes every series of events and every succession of facts becomes a necessity and a representative of national predestination. Almost everything is considered a symbol of the mysterious current of nationality, and all of us have lately read how the palaces of a great capital were conveniently proclaimed from an imperial throne to be the self-symbolizations of a nation instinctively intent on centralized unity. It is the school peculiarly in favor with modern, brilliant and not always unenlightened absolutism; for, it strikes individuality from the list of our attributes, and individuality incommodes absolutism. It is the school which strips society of its moral and therefore responsible character, and has led with us to the doctrine of manifest destiny, as if any destiny of man could be more manifest than that of doing right, above all things, and of being man indeed. The error into which this school has relapsed is not dissimilar to that which prevailed regarding ethics with the Greeks before they had clearly separated, in their minds, the laws of nature with their unbending necessity from the moral laws, and which is portrayed with fearful earnestness in the legend of Œdipus.

Closely akin in historic ethics to the theory of historical necessity is the base theory of success. We are told, and unfortunately by very many that pretend to take philosophic views, that success proves justice; that the unsuccessful cause proves by the want of success its want of right. It is a convenient theory for the tyrant; but it is forgotten that if mere prevailance of power over antagonists constitutes success, and success proves the right of the successful, the unpunished robber or the deceiver who can not be reached are justified. We are not told what length of time constitutes success. If there had been a Moniteur de Rome in the second century of our era, Christianity must have been represented as a very unsuccessful movement. Nor are we allowed to forget the strong lesson of history that no great idea, no institution of any magnitude has ever prevailed except after long struggles and unsuccessful attempts.*

The conscientious teacher must guard the young

* Connected with this error, again, is the theory of Representative Men, which seems to be in great favor at the present time, and is carried to a re-

against the blandishments of these schools; he must cultivate in the young the delight of discovering the genesis of things, which for great purposes was infused into our souls; but he must show with lasting effect, that growth in history however well traced, however

markable degree of extravagance even by men who have otherwise acquired deserved distinction. One of the most prominent philosophers of France has gone so far as to say that the leading military genius of an age is its highest representative—a position wholly at variance with history and utterly untenable by argument. The philosopher Hegel had said nearly the same thing before him. It would be absurd to say that Hannibal was the representative of his age, yet he was pre-eminently its military genius. Those are the greatest of men that are in advance of their fellow-beings and raise them up to their own height. Whom did Charlemagne represent? The question whom and what did those men represent that have been called representative men, and at what time of their lives were they such, are questions which present themselves at once at the mention of this term. An English judge who once for all has settled by his decision a point of elementary importance to individual liberty, so that his opinion and his decision now form part and parcel of the very constitution of his country, is to be considered far more a representative of the spirit of the English people than Cromwell was when he divided England into military districts, and established a government which broke down the moment he breathed his last. The greater portion of those men who are called representative men have reached their historical eminence by measures consisting in a mixture of violence, compression, and, generally, of fraud; they cannot, therefore, have represented those against whom the violence was used, and little observation is required to know that organized force or a well organized hierarchy can readily obtain a victory over a vastly greater majority that is not organized. The twenty or thirty organized men at Sing-Sing, who keep many hundred prisoners, insulated by silence, in submission, cannot be called the representative men of the penitentiary. Nor must it be forgotten that the Bad and the Criminal can be concentrated in a leader and represented by him, just as well as that which is good and substantial. Such as the idea of representative men is now floating in the minds of men, it is the result, in a great measure, of that unphilosophical coarseness which places the Palpable, the Vast and the Rapid above the silent and substantial genesis of things and ideas, thus leading to the fatal error of regarding destruction more than growth. Destruction is rapid and violent; growth is slow and silent. The naturalists have divested themselves of this barbarism.

delightful in tracing, however instructive and however enriching our associations, is not on that account alone a genesis with its own internal moral necessity, and does not on that account alone have a prescribing power for a future line of action. I have dwelt upon this subject somewhat at length, but those will pardon me who know to what almost inconceivable degree these errors are at present carried even by some men otherwise not destitute of a comprehensive grasp of mind.

If what I have said of the nourishing character inherent in the study of history is true; if history favors the growth of strong men and is cherished in turn by them, and grows upon their affection as extended experience and slowly advancing years make many objects of interest drop like leaves, one by one; if history shows us the great connection of things, that there is nothing stable but the Progressive, and that there is Alfred and Socrates, Marathon and Tours, or, if it be not quaint to express it thus, that there is the microcosm of the whole past in each of us; and if history familiarizes the mind with the idea that it is a jury whose verdict is not rendered according to the special pleadings of party dogmas, and a logic violently wrenched from truth and right—then it is obvious that in a moral, practical and intellectual point of view it is the very science for nascent citizens of a republic. There are not a few among us, who are dazzled by the despotism of a Cæsar, appearing brilliant at least at a distance—did not even Plato set, once, his hopes on Dionysius ?—or are misled by the plausible simplicity of democratic absolutism, that despotism which believes liberty simply to consist in the irresponsible power of a larger number over a smaller, for no other reason, it seems, than that ten is more than nine. All absolutism, whether monarchical or democratic, is in principle the same, and the latter always leads by short transitions to the other. We may go farther ; in all absolutism there is a strong element of communism. The theory of property which Louis the Fourteenth put forth was essentially communistic. There is no other civil liberty than institutional liberty, all else is but passing semblance and simulation. It is one of our highest duties, therefore, to foster in the young an institutional spirit, and an earnest study of history shows the inestimable value of institutions. We need not fear in our eager age and country that we may be led to an idolatry of the past—history carries sufficient preventives within itself—or to a worship of institutions simply because they are institutions. Institutions like the sons of men themselves may be wicked or good ; but it is true that ideas and feelings, however great or pure, retain a passing and meteoric character so long as they are not embodied in vital institutions, and that rights and privileges are but slender reeds so long as they are not protected and kept alive by sound and tenacious institutions; and it is equally true that an institutional spirit is fostered and invigorated by a manly study of society in the days that are gone.

A wise study of the past teaches us social analysis, and to separate the permanent and essential from the accidental and superficial, so that it becomes one of the keys by which we learn to understand better the present. History, indeed, is an admirable training in the great duty of attention and the art of observation, as in turn an earnest observation of the present is an indispensable aid to the historian. A practical life is a key with which we unlock the vaults containing the riches of the past. Many of the greatest historians in antiquity and modern times have been statesmen; and Niebuhr said that with his learning, and it was prodigious, he could not have understood Roman history, had he not been for many years a practical officer in the financial and other departments of the administration, while we all remember Gibbon's statement of himself, that the captain of the Hampshire militia was of service to the historian of Rome. This is the reason why free nations produce practical, penetrating and unravelling historians, for in them every observing citizen partakes, in a manner, of statesmanship. Free countries furnish us with daily lessons in the anatomy of states and society; they make us comprehend the reality of history. But we have dwelled sufficiently long on this branch.

As Helicon, where Clio dwelt, looked down in all its grandeur on the busy gulf and on the chaffering traffic of Corinth, so let us leave the summit and walk down to Crissa, and cross the isthmus and enter the noisy mart where the productions of men are exchanged. Sudden as the change may be, it only symbolizes reality and human life. What else is the main portion of history but a true and wise account of the high events and ruling facts which have resulted from the combined action of the elements of human life? Who does not know that national life consists in the gathered sheaves of the thousand activities of men, and that production and exchange are at all times among the elements of these activities?

Man is always an exchanging being. Exchange is one of those characteristics without which we never find man, though they may be observable only in their lowest incipiency, and with which we never find the animal, though its sagacity may have reached the highest point. As, from the hideous tattooing of the savage to our dainty adornment of the sea-cleaving prow or the creations of a Crawford, men always manifest that there is the affection of the beautiful in them—that they are æsthetical beings; or as they always show that they are religious beings, whether they prostrate themselves before a fetish or bend their knee before their true and unseen God, and the animal never, so we find man, whether Caffre, Phœnician or American, always a producing and exchanging being; and we observe that this, as all other attributes, steadily increases in intensity with advancing civilization.

There are three laws on which man's material well-being and, in a very great measure, his civilization are founded. Man is placed on this earth apparently more destitute and helpless than any other animal. Man is no finding animal—he must produce. He must produce his food, his raiment, his shelter and his comfort. He must produce his arrow and his trap, his canoe and his field, his road and his lamp.

Men are so constituted that they have far more wants, and can enjoy the satisfying of them more intensely, than other animals; and while these many wants are of a peculiar uniformity among all men, the fitness of the earth to provide for them is greatly diversified and locally restricted, so that men must produce, each more than he wants for himself, and exchange their products. All human palates are pleasantly affected by saccharine salts, so much so that the word sweet has been carried over, in all languages, into different and higher spheres, where it has ceased to be a trope and now designates the dearest and even the holiest affections. All men understand what is meant by sweet music and sweet wife, because the material pleasure whence the term is derived is universal. All men of all ages relish sugar, but those regions which produce it are readily numbered. This applies to the far greater part of all materials in constant demand among men, and it applies to the narrowest circles as to the widest. The inhabitant of the populous city does not cease to relish and stand in need of farinaceous substances though his crowded streets cannot produce grain, and the farmer who provides him with grain does not cease to stand in need of iron or oil which the town may procure for him from a distance. With what remarkable avidity the tribes of Negroland, that had never been touched even by the last points of the creeping fibres of civilization, longed for the articles lately carried thither by Barth and his companions! The brute animal has no dormant desires of this kind, and finds around itself what it stands in need of. This apparent cruelty, although a real blessing to man, deserves to be made a prominent topic in natural theology.

Lastly, the wants of men—I speak of their material and cultural wants, the latter of which are as urgent and fully as legitimate as the former—infinitely increase and are by Providence decreed to increase with advancing civilization; so that his progress necessitates intenser production and quickened exchange.

The branch which treats of the necessity, nature, and effects, the promotion and the hindrances of production, whether it be based almost exclusively on appropriation, as the fishery; or on coercing nature to furnish us with better and more abundant fruit than she is willing spontaneously to yield, as agriculture; or in fashioning, separating and combining substances which other branches of industry obtain and collect, as manufacture; or on carrying the products from the spot of production to the place of consumption; and the character which all these products acquire by exchange, as values, with the labor and services for which again products are given in exchange, this division of knowledge is called political economy—an unfit name; but it is the name, and we use it. Political economy, like every other of the new sciences, was obliged to fight its way to a fair acknowledgment, against all manners of prejudices. The introductory lecture which archbishop Whately delivered some thirty years ago, when he commenced his course on political economy in the university of Oxford, consists almost wholly of a defense of his science and an encounter with the objections then made to it on religious, moral, and almost on every ground that could be made by ingenuity, or was suggested by the misconception of its aims. Political economy fared, in this respect, like vaccination, like the taking of a nation's census, like the discontinuance of witch-trials.

The economist stands now on clearer ground. Opponents have acknowledged their errors, and the economists themselves fall no longer into the faults of the utilitarian. The economist indeed sees that the material interests of men are of the greatest importance, and that modern civilization, in all its aspects, requires an immense amount of wealth, and consequently increasing exertion and production, but he acknowledges that "what men can do the least without is not their highest need."* He knows that we are bid to pray for our daily bread, but not for bread alone, and I am glad that those who bade me teach Political Economy, assigned to me also Political Philosophy and History. They teach that the periods of national dignity and highest endeavors have sometimes been periods of want and poverty. They teach abundantly that riches and enfeebling comforts, that the flow of wine or costly tapestry, do not lead to the development of humanity, nor are its tokens; that no barbarism is coarser than

* Professor Lushington in his Inaugural Lecture, in Glasgow, quoted in Morell's Hist. and Crit. View of Specul. Phil. London, 1846.

the substitution of gross expensiveness for what is beautiful and graceful; that it is manly character, and womanly soulfulness, not gilded upholstery or fretful fashion—that it is the love of truth and justice, directness and tenacity of purpose, a love of right, of fairness and freedom, a self-sacrificing public spirit and religious sincerity, that lead nations to noble places in history; not surfeiting feasts or conventional refinement. The Babylonians have tried that road before us.

But political economy, far from teaching the hoarding of riches, shows the laws of accumulation and distribution of wealth; it shows the important truth that mankind at large can become and have become wealthier, and must steadily increase their wealth with expanding culture.

It is, nevertheless, true that here, in the most active market of our whole hemisphere, I have met, more frequently than in any other place, with an objection to political economy, on the part of those who claim for themselves the name of men of business. They often say that they alone can know anything about it, and as often ask: what is Political Economy good for? The soldier, though he may have fought in the thickest of the fight, is not on that account the best judge of the disposition, the aim, the movements, the faults or the great conceptions of a battle, nor can we call the infliction of a deep wound a profound lesson in anatomy.

What is Political Economy good for? It is like every other branch truthfully pursued, good for leading gradually nearer and nearer to the truth; for making men, in its own sphere, that is the vast sphere of exchange, what Cicero calls *mansueti*, and for clearing more and more away what may be termed the impeding and sometimes savage superstitions of trade and intercourse; it is, like every other pursuit of political science of which it is but a branch, good for sending some light, through the means of those that cultivate it as their own science, to the most distant corners, and to those who have perhaps not even heard of its name.

Let me give you two simple facts—one of commanding and historic magnitude; the other of apparent insignificance, but typical of an entire state of things, incalculably important.

Down to Adam Smith, the greatest statesmanship had always been sought for in the depression of neighboring nations. Even a Bacon considered it self-evident that the enriching of one people implies the impoverishing of another. This maxim runs through all history, Asiatic and European, down to the latter part of the last century. Then came a Scottish professor who dared to teach, in his dingy lecture-room at Edinburgh, contrary to the opinion of the whole world, that every man, even were it but for egotistic reasons, is interested in the prosperity of his neighbors; that his wealth, if it be the result of production and exchange, is not a withdrawal of money from others, and that, as with single men so with entire nations—the more prosperous the one so much the better for the other. And his teaching, like that of another professor before him—the immortal Grotius—went forth, and rose above men and nations, and statesmen and kings; it ruled their councils and led the history of our race into new channels; it bade men adopt the angels' greeting: "Peace on earth and good will towards men," as a maxim of high statesmanship and political shrewdness. Thus rules the mind; thus sways science. There is now no intercourse between civilized nations which is not tinctured by Smith and Grotius. And what I am, what you are, what every man of our race is in the middle of the nineteenth century, he owes it in part to Adam Smith, as well as to Grotius, and Aristotle, and Shakespeare, and every other leader of humanity. Let us count the years since that Scottish professor, with his common name, Smith, proclaimed his swaying truth, very simple when once pronounced; very fearful as long as unacknowledged; a very blessing when in action; and then let us answer,

What has Political Economy done for man? We habitually dilate on the effect of physical sciences, and especially on their application to the useful arts in modern times. All honor to this characteristic feature of our age—the wedlock of knowledge and labor; but it is, nevertheless, true that none of the new sciences have so deeply affected the course of human events as political economy. I am speaking as an historian and wish to assert facts. What I say is not meant as rhetorical fringe.

The other fact alluded to, is one of those historical pulsations which indicate to the touch of the inquirer, the condition of an entire living organism. When a few weeks ago the widely spread misery in the manufacturing districts of England was spoken of in the British house of lords, one that has been at the helm,* concluded his speech with an avowal that the suffering laborers who could find but half days', nay, quarter days' employment, with unreduced wants of their families, nevertheless had resorted to no violence, but on the contrary universally acknowledged that they knew full well, that a factory can not be kept working unless the master can work to a profit.

This too is very simple, almost trivial, when stated. But those who know the chronicles of the medieval

* Lord Derby, then in the opposition, and since made premier again.

cities, and of modern times down to a period which most of us recollect, know also that in all former days the distressed laborer would first of all have resorted to a still greater increase of distress, by violence and destruction. The first feeling of uninstructed man, produced by suffering, is vengeance, and that vengeance is wreaked on the nearest object or person; as animals bite, when in pain, what is nearest within reach. What has wrought this change? Who, or what has restrained our own sorely distressed population from blind violence, even though unwise words were officially addressed to them, when under similar circumstances in the times of free Florence or Cologne there would have been a sanguinary rising of the "wool-weavers," if it is not a sounder knowledge and a correcter feeling regarding the relations of wealth, of capital and labor, which in spite of the absurdities of communism has penetrated in some degree all layers of society? And which is the source whence this tempering knowledge has welled forth, if not Political Economy?

True indeed, we are told that economists do not agree; some are for protection, some for free trade. But are physicians agreed? And is there no science and art of medicine? Are theologians agreed? Are the cultivators of any branch of knowledge fully agreed, and are all the beneficial effects of the sciences debarred by this disagreement of their followers? But, however important at certain periods the difference between protectionists and free-traders may be, it touches, after all, but a small portion of the bulk of truth taught by Political Economy, and I believe that there is a greater uniformity of opinion, and a more essential agreement among the prominent scholars of this science, than among those of others excepting, as a matter of course, the mathematics.

If it is now generally acknowledged that Political Economy ought not to be omitted in a course of superior education, all the reasons apply with greater force to that branch which treats of the relations of man as a jural being—as citizen, and most especially so in our own country, where individual political liberty is enjoyed in a degree in which it has never been enjoyed before. Nowhere is political action carried to a greater intensity, and nowhere is the calming effect of an earnest and scientific treatment of politics more necessary. In few countries is man more exposed to the danger of being carried away to the worship of false political gods and to the idolatry of party, than in our land, and nowhere is it more necessary to show to the young the landmarks of political truth, and the essential character of civil liberty—the grave and binding duties that man imposes upon himself when he proudly assumes self-government. Nowhere seem to be so many persons acting on the supposition that we differ from all other men, and that the same deviations will not produce the same calamities, and nowhere does it seem to be more necessary to teach what might well be called political physiology and political pathology. In no sphere of action does it seem to me more necessary than in politics, to teach and impress the truth that "logic without reason is a fearful thing." Aristotle said: The fellest of things is armed injustice; History knows a feller thing —impassioned reasoning without a pure heart in him that has power in a free country—the poisoning of the well of political truth itself. Every youth ought to enter the practical life of the citizen, and every citizen ought to remain through life, deeply impressed with the conviction that, as Vauvernague very nobly said, "great thoughts come from the heart," so great politics come from sincere patriotism, and that without candid and intelligent public spirit, parties without which no liberty can exist, will raise themselves into ends and objects instead of remaining mere means. And when the words party, party consistency and party honor are substituted for the word Country, and, as Thucydides has it, when parties use, each its own language, and men cease to understand one another, a country soon falls into that state in which a court of justice would find itself where wrangling pleaders should do their work without the tempering, guiding judge—that state of dissolution which is the next step to entire disintegration. Providence has no special laws for special countries, and it is not only true what Talleyrand said: *Tout arrive;* but everything happens over again. There is no truth, short of the multiplication table, that, at some time or other, is not drawn into doubt again, and must be re-asserted and re-proved.

One of the means to insure the difficult existence of liberty—far more difficult than that of absolutism, because of an infinitely more delicate organization—is the earnest bringing up of the young in the path of political truth and justice, the necessity of which is increased by the reflection that in our period of large cities, man has to solve, for the first time in history, the problem of making a high degree of general and individual liberty compatible with populous cities. It is one of the highest problems of our race, which cannot yet be said to have been solved.

Political philosophy is a branch of knowledge that ought not only to be taught in colleges; its fundamental truths ought to be ingrained in the minds of every one that helps to crowd your public schools. Is it objected that political philosophy ranges too high for boyish intellects? What ranges higher, what is of so spiritual a character as christianity?

But this has not prevented the church, at any period of her existence, from putting catechisms of a few pages into the hands of boys and girls, so that they could read.

We have, however, direct authority for what has been advanced. The Romans in their best period made every school-boy learn by heart the XII Tables, and the XII Tables were the catechism of Roman public and private law, of their constitution and of the proud Jus Quiritium, that led the Roman citizen to pronounce so confidently, as a *vox et invocatio*, his Civis Romanus sum in the most distant corners of the land, and which the captive apostle collectedly asserted twice before the provincial officers. Cicero says that when he was a boy, he learned the XII Tables *ut carmen necessarium*, like an indispensable formulary, a political breviary, and deplores that at the time when he was composing his treatise on the laws, in which he mentions the fact, the practice was falling into disuse. Rome was fast drifting to Cæsarean absolutism; what use was there any longer for a knowledge of fundamental principles?

The Romans were not visionary; they were no theorists; no logical symmetry or love of system ever prevented them from being straightforward and even stern practical men. They were men of singular directness of purpose and language. Abstraction did not suit them well. Those Romans, who loved law and delighted in rearing institutions and building high roads and aqueducts; who could not only conquer, but could hold fast to, and fashion what they had conquered, and who strewed municipalities over their conquests, which, after centuries, became the germs of a new political civilization; who reared a system of laws which conquered the west and their own conquerors, when the Roman sword had become dull; and who impressed, even through the lapse of ages, a practical spirit on the Latin Church, which visibly distinguishes it from the Greek; those Romans who declared their own citizens with all the Jus Romanum on them, when once enrolled, the slaves of the general, and subjected them to a merciless whip of iron chains; those Romans who could make foreign kings assiduous subjects, and foreign hordes fight well by the side of their own veterans, and who could be dispassionately cruel when they thought that cruelty was useful; those Romans who were practical if there ever was a practical people, bade their schoolmaster to drive the XII Tables into the stubborn minds of the little fellows who, in their turn, were to become the ruling citizens of the ruling commonwealth, and we know, from sculptural and written records, in prose and metre, that the magistral means in teaching that carmen necessarium was not always applied to the head alone.

Let us pass to another authority, though it require a historic bound—to John Milton, whose name is high among the names of men, as that of Rome is great among the states of the earth. Milton who wrote as clear and direct prose, as he sang lofty poetry, who was one of the first and best writers on the liberty of the press against his own party, and who consciously and readily sacrificed his very eyesight to his country—Milton says, in his paper on Education, dedicated to Master Hartlib,* that, after having

* Mr. Evert A. Duyckinck, of this city, whom, while writing out this address, I had asked what he knew of "Master Hartlib," obligingly replied by a note, of which I may be permitted to give the following extract:

"In D'Israeli's Curiosities of Literature, Hartlib is called a Pole. Thomas Wharton, in a note in his edition of Milton's Minor Poems, says Hartlib was a native of Holland, and came into England about the year 1640. Hartlib himself tells us in a letter, dated 1660, (reprinted in Egerton Brydges' Curiosa Literaria, III. 54) that his father was a Polish merchant who founded a church in Pomania, and, when the Jesuits prevailed in Poland, removed to Elbing, to which place his (Samuel Hartlib's) grandfather brought the English company of merchants from Dantzic. It would appear that Hartlib was born at Elbing, for he speaks of his father marrying a third wife (H.'s mother) after the removal from Poland proper, which third wife would appear to have been an English woman. Hartlib speaks of his family being 'of a very ancient extraction in the German empire, there having been ten brethren of the name of Hartlib, some of them Privy Councillors to the Emperor.' Hartlib's mercantile life, I suppose, brought him to England. He was a reformer in Church matters, and became attached to the Parliament. 'Hartlib,' says Wharton, 'took great pains to frame a new system of education, answerable to the perfection and purity of the new commonwealth.' Milton addressed his Treatise on Education to him about 1650. In 1662 Hartlib petitioned Parliament for relief, stating that he had been thirty years and upwards serving the state and specially setting forth the 'erecting a little academy for the education taught sundry other branches in a boy's education, "the next removal must be to the study of politics, to know the beginning, end and reasons of political societies, that they (the learners) may not, in a dangerous fit of the commonwealth, be such poor, shaken, uncertain reeds, of such a tottering conscience, as many of our great counsellors have lately shown themselves, but steadfast pillars of the state." This pregnant passage ought not to have been written in vain.

I could multiply authorities of antiquity and modern times, but is not, Rome and Milton, strong enough?

A complete course of political philosophy, to which every course, whether in a college or a university, ought to approximate, as time and circumstances permit, should wind its way through the large field of political science somewhat in the following manner.

of the gentry of this nation, to advance piety, learning, morality and other exercises of industry, not usual then in common schools.' His other services were 'correspondence with the chief of note of foreign parts,' 'collecting MSS. in all the parts of learning,' printing 'the best experiments of industry in Husbandry and Manufactures,' relieving 'poor, distressed scholars, both foreigners and of this nation.'"

So far the extract from Mr. Duyckinck's letter. Hartlib was no doubt a German by extraction and education, and represents a type of men peculiar to the reformation, and of great importance in the cause of advancing humanity. Milton must have felt great regard for this foreigner, but Milton had too enlightened a mind, and had learned too much in foreign parts, ever to allow a narrowing and provincial self-complacency to become a substitute for enlarging and unselfish patriotism.

We must start from the pregnant fact that each man is made an individual and a social being, and that his whole humanity with all its attributes, moral, religious, emotional, mental, cultural and industrial, is decreed forever to revolve between the two poles of individualism and socialism, taking the latter term in its strictly philosophical adaptation. Man's moral individualism and the sovereign necessity of his living in society, or the fact that humanity and society are two ideas that cannot even be conceived of, the one without the other, lead to the twin ideas of Right and Duty. Political science dwells upon this most important elementary truth, that the idea of right cannot be philosophically stated without the idea of obligation, nor that of duty without that of right, and it must show how calamitous every attempt has proved to separate them; how debasing a thing obligation becomes without corresponding rights, and how withering rights and privileges become to the hand that wields the power and to the fellow-being over whom it sways, if separated from corresponding duty and obligation.

Right and duty are twin brothers; they are like the two electric flames appearing at the yard-arms in the Mediterranean, and were called by the ancient mariners Castor and Pollux. When both are visible, a fair and pleasant course is expected; but one alone portends stormy mischief. An instinctive acknowledgment of this truth makes us repeat with pleasure to this day the old French maxim, *Noblesse oblige*, whatever annotations history may have to tell of its disregard.*

That philosopher, whom Dante calls *il maestro di color che sanno*, and whom our science gratefully acknowledges as its own founder, says that man is by nature a political animal. He saw that man can not divest himself of the State. Society, no matter in how rudimental a condition, always exists, and society considered with reference to rights and duties, to rules to be obeyed, and to privileges to be protected, to those that ordain, and those that comply, is the political state. Government was never voted into existence, and the state originates every day anew in the family. God coerces man into society, and necessitates the growth of government by that divinely simple law, which has been alluded to before, and consists in making the young of man depend upon the parents for years after the period of lactation has ceased. As men and society advance, the greatest of institutions—the State—increases in inten-

* In this sense at least Noblesse oblige was often taken, that feudal privileges over feudal subjects involved obligations to them, although it meant originally the obligations due to him who bestowed the nobility.

sity of action, and when humanity falters back, the State, like the function of a diseased organ, becomes sluggish or acts with ruinous feverishness. In this twinship of right and duty lies the embryonic genesis of liberty, and at the same time the distinction between sincere and seasoned civil liberty, and the wild and one-sided privilege of one man or a class; or the fantastic equality of all in point of rights without the steadying pendulum of mutual obligation.

This leads us to that division which I have called elsewhere Political Ethics, in which the teacher will not fail to use his best efforts, when he discourses on patriotism—that ennobling virtue which at times has been derided, at other times declared incompatible with true philosophy or with pure religion. He will not teach that idolatrous patriotism which inscribes on its banner, Our country, right or wrong, but that heightened public spirit, which loves and honors father and mother, and neighbors, and country; which makes us deeply feel for our country's glory and its faults; makes us willing to die, and, what is often far more difficult, to live for it; that patriotism which is consistent with St. Paul's command: Honor all men, and which can say with Montesquieu, "If I knew anything useful to my country but prejudicial to Europe or mankind, I should consider it as a crime;" that sentiment which made the Athenians reject the secret of Themistocles, because Aristides declared it very useful to Athens, but very injurious to Sparta and to the other Greeks. The christian citizen can say with Tertullian, *Civitas nostra totus mundus*, and abhors that patriotism which is at best bloated provincialism, but he knows, too, that that society is doomed to certain abasement in which the indifference of the *blasé* is permitted to debilitate and demoralize public sentiment. The patriotism of which we stand as much in need as the ancients, is neither an amiable weakness, nor the Hellenic pride. It is a positive virtue demanded of every moral man. It is the fervent love of our own country, but not hatred of others, nor blindness to our faults and to the rights or superiorities of our neighbors.

We now approach that branch of our science which adds, to the knowledge of the "end and reasons of political societies," the discussion of the means by which man endeavors to obtain the end or ought to obtain it; in one word, to the science of government, and a knowledge of governments which exist and have existed. The "end and reasons of political societies" involve the main discussion of the object of the State, as it is more clearly discerned with advancing civilization, the relation of the State to the family, its duties to the individual, and the necessary limits of its power. Protection, in the highest sense of the word,* both of society, as a whole, and of the component individuals, as such, without interference, and free from intermeddling, is the great object of the civilized State, or the State of freemen. To this portion of our science belong the great topics of the rights as well as the dependence of the individual citizen, of the woman and the child; of primordial rights and the admissibility or violence of slavery, which, throughout the whole course of history wherever it has been introduced, has been a deciduous institution. The reflection on the duties of the State comprehends the important subjects of the necessity of public education (the common school for those who are deprived of means, or destitute of the desire to be educated; and the university, which lies beyond the capacity of private means); of the support of those who cannot support themselves (the pauper, and the poor orphans and sick); of intercommunication and intercommunion (the road and the mail); of the promotion of taste and the fine arts, and the public support of religion, or the abstaining from it; and the duty of settling conflicting claims, and of punishing

* That I do not mean by this material protection only, but the protection of all interests, the highest no less so than the common ones, of society as a unit, as well as of the individual human being, will be well known to the reader of my Political Ethics. I do by no means restrict the meaning of Protection to personal security, nor do I mean by this term something that amounts to the protection of an interest in one person to the injury of others.

those that infringe the common rules of action, with the science and art of rightful and sensible punition, or, as I have ventured to call this branch, of penology.

The comprehensive apparatus by which all these objects, more or less dimly seen, according to the existing stage of civil progress, are intended to be obtained, and by which a political society evolves its laws, is called government. I generally give at this stage a classification of all governments, in the present time or in the past, according to the main principles on which they rest. This naturally leads to three topics, the corresponding ones of which, in some other sciences, form but important illustrations or constitute a certain amount of interesting knowledge, but which in our science constitute part and parcel of the branch itself. I mean a historical survey of all governments and systems of law, Asiatic or European; a survey of all political literature as represented by its prominent authors, from Aristotle and Plato, or from the Hindoo Menu, down to St. Simon or Calhoun—a portion of the science which necessarily includes many historians and theologians on the one hand, such as Mariana, De Soto and Machiavelli, and on the other hand statesmen that have poured forth wisdom or criminal theories in public speech, Demosthenes or Webster, Chatham, Burke, Mirabeau or Robespierre and St. Juste. And lastly, I mean that division of our science which indeed is, properly, a subdivision of the latter, but sufficiently important and instructive to be treated separately—a survey of those model states which political philosophers have from time to time imagined, and which we now call Utopias, from Plato's Atlantis to Thomas More's Utopia, Campanella's Civitas Solis or Harrington's Oceana to our socialists, or Shelley's and Coleridge's imaginings and the hallucinations of Comte. They are growing rarer and, probably, will in time wholly cease. Superior minds, at any rate, could feel stimulated to conceive of so-called philosophical republics, in ages only when everything existing in a definite form—languages, mythologies, agriculture and governments—was ascribed to a correspondingly definite invention, or, at times, to an equally definite inspiration, and when society was not clearly conceived to be a continuity; when far less attention was paid to the idea of progress, which is a succession of advancing steps, and to the historic genesis of institutions; and when the truth was not broadly acknowledged that civilization, whether political or not, cannot divest itself of its accumulative and progressive character.

This Utopiology, if you permit me the name, will include those attempts at introducing, by sudden and volcanic action, entirely new governments resulting from some fanatical theory, such as the commonwealth of the anabaptists in Germany, or the attempts of carrying out Rousseau's equalitarian hatred of representative government, by Marat and Callot d'Herbois. They have all been brief and bloody.

When the teacher of political philosophy discourses on the first of these three divisions he will not omit to dwell on the communal governments and the later almost universal despotism, of Asia, which reduces the subject, both as to property and life, to a tenant at will; he will dwell on the type of the city-state, prevailing in Greek and Roman antiquity, and the strong admixture of communism in those states, especially in the Greek; he will show how that religion, whose founder proclaimed that his kingdom is not of this world, nevertheless affected all political organization far more than aught else has done, because, more than anything else, it affected the inner man, and that, in one respect, it intensified individualism, for it exalted the individual moral character and responsibility. Individual duties and individual rights received greater importance, and christianity leveled all men before an omniscient Judge and a common Father. From the time when the worshiped emperor of Rome decreed that the christians, then confounded with the Jews, should depart from Italy, because, as Suetonius says, they were *Christo impulsore tumultuantes*, the Romans perceived that there was that in the christian which made him bow before a higher authority than that of the Cæsars. "Christ impelling them, they are disturbers"—yet they obeyed the law, as Pliny, the governor, writes to his friend and emperor, only they could not be induced to strew the sacred meal on the altar of Jove, and christianity wrought on in the breasts of men, until Julian loses the battle, and, as tradition at least says, exclaims in dying: "Oh Galilean, thou hast conquered!"

The teacher will dwell on that type of government which succeeded and is the opposite to the ancient city-state—the feudal system with its graduated and subdivided allegiance; and he will show how at last the period of nationalization arrived for governments and languages, and national governments, with direct and uniform allegiance, at last developed selves and became the accompaniments of modern civilization; when real states were formed, compact governments extending over large territories. The ancients had but one word for state and city; the mediæval government is justly called a mere system (the feudal system); the moderns have states, whether unitary or confederated does not affect this point.

When an account is given of the imaginary governments, which the greater or lesser philosophers have constructed as ideal polities, attention must be directed to the striking fact that all Utopists, from Plato to our times, have been more or less communists, making war upon money, although so shrewd and wise a man as Thomas More was among them; and that most of these writers, even Campanella, though a priest of the catholic church, and all societies in which communism has been carried out to any extent, have made light of monogamic wedlock, or have openly proclaimed the community or a plurality of wives.*

* Auguste Comte, who was generally considered the most serious and most able atheist, yet known in the annals of science, as long as his Positive Philosophy was the only work that attracted attention, makes one of the exceptions. In his Catechism of Positive Religion, which belongs to the Utopian literature, proclaiming the regeneration and the reconstruction of all human society, and covering it with the ægis of a paper-system rubricked according to a priestly socialistic Cæsarism, nevertheless acknowledges monogamy, and individual property in a considerable degree. The work, however, amply makes up for these omissions, by an incredible amount of inane vagaries, self-contradictions and that apotheosis of absolutism, "organizing" all things and allowing inherent life nowhere, which is the idol of Gallican sociologists, as the fallen Romans burnt incense to the images of their emperors even while living, or rather as long as they lived; for, so soon as the emperor was dead, his memory was often senatorially cursed, and his images were decreed to be broken. Power was the only thing left, when the introduction of the many thousands of gods, from the conquered countries, neutralized all sense of religion, and power was worshiped accordingly. The Suetoniana of the nineteenth century are not wholly dissimilar.

Nothing has probably ever shown so strikingly the inherent religious character of man as Comte's apotheosis of atheism, and his whole "catechism," sprinkled as it is with prayers to the "supreme being," which being, to be sure, is void of being and cannot, therefore, very well be possessed of supremacy.

From time to time great men have declared what they considered the greatest of evils. Aristotle says, "The fellest of things is armed injustice." Bacon declares that the greatest of evils is the apotheosis of error; but, somehow, men seem always to contrive to prove that there may be still greater evils.

We have our protestant counterpart to Campanella in the Rev. Martin Madan, the author of Thelypthora, a Defense of a Plurality of Wives. Hostility to individualism in property has generally been accompanied by a hostility to exclusive wedlock, in antiquity and modern times, and I believe I am not wrong when I add, very often by a leaning to pantheism, in the sphere of religion. But the Utopists are not the only communists. Paley, who would have shrunk from being called a communist, nevertheless explains individual property on the mere ground of his "expediency" and in a manner which the avowed communists of our times—Quinesset and Proudhon—have been willing to accept, only they differ as to the expediency, and why not differ on that? Paley and the larger portion of modern publicists maintained, and even Webster asserted on a solemn occasion,* that property is the creature of government. But government is the agent of society, so that, if the same society should see fit to change the order of things, and to undo its own doing, no objection can be made on the ground of right and justice. Rousseau says, indeed, that the first fence erected to separate land from the common stock, brought misery

* It was the perusal of this assertion by Mr. Webster, in a speech in Ohio, in 1828, which first led the author to reflections which were ultimately given in his Essays on Labor and Property. He totally denies that property is the creature of government.

upon men, and Proudhon *formulated* this idea when he said: Property is theft; but the point of starting is common to all.

The radical error of the communist consists in his exclusive acknowledgment of the principle of socialism, and that he endeavors to apply it even to that which has its very origin and being in individualism—to property. Man can not exist without producing; production always presupposes appropriation; both are essentially individual, and where appropriation consists in occupation by a society as a unit, this is no less exclusive or individual property, with reference to all other societies, than the property held by a single man. The communist does not seem to see the absurdity of demanding common property for all men in France, upon what he considers philosophic grounds, yet excluding the rest of mankind from that property. The radical error of the individualist, on the other hand, is, that he wholly disavows the principle of socialism, and, generally, reasons on the unstable and shaking ground of expediency alone. He forgets that both, individualism and socialism, are true and ever-active principles, and that the very idea of the state implies both; for, the state is a society, and a society consists of individuals who never lose their individual character, but are united by common bonds, interests, organizations and a common continuity. A society is not represented by a mass of iron in which the original particles of the ore have lost all separate existence by refinement and smelting; nor is it represented by a crowd of units accidentally huddled together. It is on the principle of socialism alone that it can be explained why I may be forced, and ought to be so, to pay my share toward the war which I may loathe, but upon which my state, my society has resolved. How will you explain that charity is no longer left wholly to depend upon individual piety, but that the government takes part of my property in the shape of a poor-tax, to support the indigent? or how is the potent right of roads to be explained? that I must pay toward common education when I may educate my children in a private school or may have none at all to be educated? or toward a scientific expedition, or to support the administration of justice, when I may not have had a single law-suit or when I might think it more convenient to return to the primitive age of private revenge? On what principle do you prohibit infamous books? Why must I bear the folly of my legislators or submit to the consequences of a crude diplomatist? Why are we proud of the willing submission of the minority after a passionately-contested presidential election? The principle of socialism is interwoven with our whole existence; for, it is a social existence.

How, again, can we explain the very idea of rights, the protection of man, all the contents of all the bills of rights—the liberty of the press or communion, the freedom of worship or the right we have to slay the sheriff that breaks into our house with an illegal warrant, if not on the ground of individualism? All taxation is founded on socialism, inasmuch as society takes by force, actual or threatened, part of my own, and on individualism, because it is proportioned according to the capacity of the individual to pay and takes a lawful portion only. When the Athenian council decreed a liturgy, there was socialism indeed pretty strongly prevailing. The principle of individualism is everywhere, for our existence is, also, an individual one. We shudder instinctively at the idea of losing our individuality, and our religion teaches that we must take it with us beyond the limits of time. Even a heathen, a Hindoo law-giver, said long before our era: "Single is each man born; single he dieth; single he receiveth the reward of his good, and single the punishment for his evil deeds."

The two principles of humanity, individualism and socialism, show themselves from the very beginning in their incipient pulsations, and as mankind advance they become more and more distinct and assume more and more their legitimate spheres. Individualism is far more distinct with us than in antiquity, in property and in the rights of man, with all that flows from them; and socialism is far more clearly developed with us than with the Greeks or Romans, in primary education, charity, intercommunion by the liberty of the press or the mail, the punitary systems, sanitary measures, public justice and the many spheres in which the united private wants have been raised to public interests, and often passed even into the sphere of international law. Christianity, which, historically speaking, is a co-efficient of the highest power of nearly all the elements of humanity and civilization, has had an intensifying effect on individualism as well as on socialism. There is, perhaps, no more striking instance of a higher degree of individualism and socialism developed at the same time, than in the administration of penal justice, which always begins with private revenge and gradually becomes public justice, when the government obliges every one to pay toward the punishment of a person that has directly injured only one other individual. Yet individualism is more developed in this advanced administration of justice, inasmuch as it always pronounces clearer and clearer, and more and more precautions are taken, that the individual wrong-doer alone shall suffer. There is no atonement demanded, as was the case with the Greeks, but plain punishment for a proved wrong, so that, if the crime is proved but not the criminal, we do not demand, on the ground of socialism, the suffering of some one, which the Greeks frequently did.

Act on individualism alone, and you would reduce society to a mere crowd of egotistical units, far below the busy but peaceful inmates of the ant-hill; act on socialism alone, and you reduce society to loathsome despotism, in which individuals would be distinguished by a mere number, as the inmates of Sing Sing. Despotism, of whatever name, is the most equalitarian government. The communist forgets that communism in property, as far as it can exist in reality, is a characteristic feature of low barbarism. Herodotus tells us what we find with existing savages. Mine and Thine in property and marriage is but dimly known by them. The communist wants to "organize," as he calls it, but in fact to disindividualize everything, even effort and labor, and a garden of the times of Louis XV., in which the ruthless shears have cramped and crippled every tree into a slavish uniformity, seems to delight his eye more than a high forest, with its organic life and freedom. Hobbes, who, two centuries ago, passed through the whole theory of all-absorbing power conveyed to one man by popular compact, which we now meet with once more in French Cæsarism, defined religion as that superstition which is established by government, and we recollect how closely allied all despotism is to communism. The highest liberty—that civil freedom which protects individual humanity in the highest degree, and at the same time provides society with the safest and healthiest organism through which it obtains its social ends of protection and historic position—may not inaptly be said to consist in a due separation and conjunction of individualism and socialism.*

One more remark. It is a striking fact that the old adage, all extremes meet, has been illustrated by none more forcibly than by the socialists; for the most enthusiastic socialists of France, America and Germany have actually come to the conclusion, that there need be and ought to be no government at all among men truly free, except, indeed, as one of

* It is for these reasons that the new term sociology seems to be inappropriate. Years ago it suggested itself to the author, when he desired to find a term more comprehensive and more compact than that of political philosophy, but he soon discarded it. If those French writers adopt it, in whose theories the idea of society absorbs almost all individualism, it is consistent. With them society, or the government which is its agent, whether monarchical or republican, is expected and demanded to provide for everything, to organize all relations, and to do all things that can possibly be done by the government; but it is to be regretted that men like Lord Brougham have adopted the term. The national society ought not to be the all-absorbing one, nor is the jural society the only important society to which the individual of our race belongs. We belong to societies of great importance, which are narrower than the State, and to others which extend far beyond it, as is sufficiently shown by the religious society or church, the œconomical society or society of production and exchange, the society of comity, the society of letters and science (for instance, in Germany or that which covers England and the United States), and the international society embracing all the Cis-Caucasian people.

our own most visionary socialists naïvely adds, for roads and some such things. For them Aristotle discovered in vain that: Man is by nature a political animal. "Leave them and pass on."

The political philosopher will now take in hand, as a separate topic, our own polity and political existence; and this will lead to our great theme, to a manly discussion of Civil Liberty and Self-Government. We are here in the peristyle of a vast temple, and I dare not enter it with you at present, for fear that all the altars and statues and votive tablets of humanity, with all the marbled records of high martyrdom and sanguinary errors, would detain us far beyond the midnight hour. It is our American theme, and we, above all men, are called upon to know it well, with all the aspirations, all the duties and precious privileges, all the struggles, achievements, dangers and errors, all the pride and humiliation, the checks and impulses, the law and untrameled action, the blessings and the blood, the great realities, the mimicry and licentiousness, the generous sacrifices and the self-seeking, with all these memories and actualities—all wound up in the memory of that one word Liberty.

And now the student will be prepared to enter upon that branch which is the glory of our race in modern history, and possibly the greatest achievement of combined judgment and justice, acting under the genial light of culture and religion—on International Law, that law which has gathered even the ocean under its fold. The ancients knew it not in their best time; and life and property, once having left the shore, were out of the pale of law and justice. Even down to our Columbus, the mariner stood by the helm with his sword, and watched the compass in armor.

Political science treats of man in his most important earthly phase; the State is the institution which has to protect or to check all his endeavors, and, in turn, reflects them. It is natural, therefore, that a thorough course of this branch should become, in a great measure, a delineation of the history of civilization, with all the undulations of humanity, from that loose condition of men in which Barth found many of our fellow-beings in Central Africa, to our own accumulated civilization, which is like a rich tapestry, the main threads of which are Grecian intellectuality, christian morality and trans-mundane thought, Roman law and institutionality, and Teutonic individual independence, especially developed in Anglican liberty and self-government.

Need I add that the student, having passed through these fields and having viewed these regions, will be the better prepared for the grave purposes for which this country destines him, and as a partner in the great commonwealth of self-government? If not, then strike these sciences from your catalogue. It is true, indeed, that the scholar is no consecrated priest of knowledge, if he does not love it for the sake of knowledge. And this is even important in a practical point of view; for all knowledge, to be usefully applied, must be far in advance of its application. It is like the sun, which, we are told, causes the plant to grow when he has already sunk below the horizon; yet I acknowledge without reserve, for all public instruction and all education, the token which I am in the habit of taking into every lecture room of mine, to impress it ever anew on my mind and on that of my hearers, that we teach and learn:

NON SCHOLÆ SED VITÆ,* VITÆ UTRIQUE.

* Seneca.

MATHEMATICS:

INAUGURAL ADDRESS

OF

CHARLES DAVIES, LL. D.,

PROFESSOR OF MATHEMATICS IN COLUMBIA COLLEGE,

ON

The Nature, Language, and Uses

OF

MATHEMATICAL SCIENCE,

February 11th, 1858.

ADDRESS.

THE first, and surely the most difficult duty assigned to me by the Board of Trustees, is that of explaining to a popular audience the nature of Mathematical Science—the forms of its language—its uses as a means of mental training and development—its value as the true basis of the practical—the sources of knowledge which it opens to the mind and the place which it should occupy in a justly balanced system of Collegiate instruction.

The term Mathematics, as used by the ancients, embraced every known Science and was also applicable to all other branches of Knowledge. Subsequently, it was restricted to those more difficult subjects which require continuous attention, severe study, patient investigation and exact reasoning; and such subjects were called Disciplinal, or Mathematical.

Mathematics, as a science, is conversant about the laws of Numbers and Space. The two abstract quantities, Number and Space, are the only subjects of Mathematical Science. The laws which are evolved in the processes employed in searching out the elements of these abstract quantities, in discussing their relations, and in framing a proper language by means of which these relations can be recorded and a knowledge of them communicated, constitute the Science of Mathematics. The faculties of the mind chiefly employed in the cultivation of this Science are simply, the apprehension, the judgment and the reasoning faculty.

The term quantity, applicable both to number and space, embraces but eight classes of units: 1st, Abstract Units; 2d, Units of Currency; 3d, Units of Length; 4th, Units of Surface; 5th, Units of Volume; 6th, Units of Weight; 7th, Units of Time; and 8th, Units of Angular Measure.

The laws which make up the Science of Mathematics are established in a series of logical propositions, deduced from a few self-evident notions of these unities, which are all referred to number and space. All the definitions and axioms, and all the truths deduced from them, by processes of reasoning, are therefore traceable to these two sources.

In mathematics, names imply the existence of the things which they name, and the definitions of those names express attributes of the things. Hence, all definitions do, in fact, rest on the intuitive inference that things corresponding to the words defined have a conceivable existence as subjects of thought, and do, or may have, proximatively, an actual existence. Every definition of this class is a tacit assumption of some proposition, which is expressed by means of the definition, and which gives to such definition its importance.

The axioms of Geometry are intuitive inductions; that is, they are perfectly conceived by a single process of the mind, without the intervention of other ideas, the moment the facts on which they depend are apprehended. When we say, "A whole is equal to the sum of all its parts," or, "A whole is greater than any of its parts," the mind immediately refers to a single thing, divided into parts; it then compares the whole thing with all its parts, or the whole thing with some of its parts; and then infers, by a process of generalization, that what is true of one thing and its parts is also true of every other thing and its parts: so that these axioms, however self-evident, are still generalized propositions, and so far of the inductive kind, that, independently of experience, they would not present themselves spontaneously to the mind.

The pure mathematics being based on definitions and axioms, as premises, all its truths are established by processes of deductive reasoning; hence, it is purely a deductive science. If all the connections between the minor and major premises were obvious to the senses, or as evident as the truth, "A whole is equal to the sum of all its parts," there would be no necessity for trains of reasoning, and deductive science would not exist. Trains of reasoning are necessary for extending the definitions and axioms to new cases; and there is no logical test of truth, in the whole range of mathematical science, except in the conformity of the conclusions to the definitions and axioms, or to such known principles as may have been established from them.

Language is a collection of all the signs of thought by means of which we express our ideas and their relations. The language of mathematics is mixed. It is composed partly of symbols, which have a precise and known signification, and partly of words borrowed from our common language. The symbols, although arbitrary marks, are, nevertheless, entirely general in their signification, as signs and instruments of thought, and when the sense in which they are used is once fixed, by definition, they always retain the same meaning throughout the same process. The meaning of the words taken from our common vocabulary is often modified and sometimes entirely changed, when transferred to the language of science. They are then used in a particular sense and are said to have a technical signification.

There are three principal forms, or dialects of the

Mathematical language: the language of Number, of which the elementary symbols are the ten figures: the language of Geometry, of which the elements are the right line and the curve; and the more comprehensive language of Analysis, in which the quantities considered, whether numerical, concrete, or appertaining to space, are represented by letters of the alphabet. These three forms of language are the basis of classification, and the science of mathematics is divided into three corresponding parts: Arithmetic, Geometry, and Analysis.

The alphabet of the Arithmetical language contains ten characters, called figures, each of which has a name, and when standing by itself indicates as many things as that name denotes. There are but three combinations of these characters—the first is formed by writing them in rows—the second by writing some of them over or under others—and the third, by means of the decimal point. This language, having ten elements and three combinations, is more simple, more minute, and more exact than any other known form of expressing our thoughts. It records all the daily transactions of the world, involving number and quantity. The yearly income—the accumulation of property—the balance sheets of mercantile enterprise are all expressed in numbers, and may be written in figures. These ten little characters are not only the sleepless sentinels of trade and commerce, but they also make known all the practical results of scientific labor.

The language of Geometry is pictorial, and has but two elements, the straight line and curve. The combinations of these simple elements give every form and variety of the geometrical language. Distance, surface, volume and angle, are names denoting portions of space. Under these four names every part of space, in form, extent and dimension, is represented to the mind by means of the straight line and curve. This language is both simple and comprehensive. The shortest distance—the curve of grace and beauty—the smooth surface and the rugged boundary are alike amenable to its laws. It presents to the mind, through the eye, the forms and relative magnitudes of all the heavenly bodies, and, also, of the most minute and delicate objects that are revealed by the microscope. It is the connecting link between theoretical and practical knowledge in the mechanic arts, and the only language in which science speaks to labor. All the works of Architecture, Sculpture and Painting, are but images of the imagination until they assume the geometrical forms.

The language of analysis is more comprehensive than the language of figures or the pictorial language of geometry; indeed, it embraces them both.

Its elements are the leading and final letters of the alphabet, and a few arbitrary signs. The combinations of these elements are few in number and simple in form; and from these humble sources are derived the fruitful language of analytical science. This language is minute, suggestive, certain, general and comprehensive. It will express every property and relation of number—every form which the imagination has given to space—every moment of time which has elapsed since hours began to be numbered—and every motion which has taken place since matter began to move. One or the other of these three forms of mathematical language is in daily use in every part of the world, and especially so in every place where science is employed to guide the hand of labor—to investigate the laws of matter—or to enlarge the boundaries of knowledge.

Of all mysteries, none is greater than the mystery of language. The invisible essence which we call mind, holds no communion with other minds, except through the double system of signs, the language of the eye and the language of the ear. Destroy the power of language, and the lights of knowledge would be extinguished. Man would live only in the present. The past and the future would be equally beyond his reach. Through language we look back over the records of the past, and trace the progress of our race through all its vicissitudes and changes from the very cradle of Creation. The wisdom of philosophy—the power of eloquence—the graces of rhetoric and the inspirations of poetry, thus become the property of every age and the common heritage of mankind. Scientific language reaches even over a wider field. The laws of the material world are the truths which it records, and the thoughts of God, manifested in all the works of the visible creation, are the treasures of its literature.

The first step in mental training is to furnish the mind with clear and distinct ideas, with settled names; each idea and its name being so associated that the one shall always suggest the other. The ideas which make up our knowledge of mathematics fulfill exactly these requirements. They are expressed in a fixed, definite and certain language, which in all its elementary forms may be illustrated by images or pictures, clear and distinct in their outlines, and having names which suggest at once their characteristics and properties.

By means of visible representations of lines, surfaces and volumes, the mind contemplates the abstract, as it were, with a thinking eye. Form, figure, distance, space, and the laws relating to them, are thus rendered familiar through the visibility of pictorial representations. This pictorial language imparts a deep interest, both from its certainty and its influence on the imagination—it attracts and animates the minds of the young, and gradually prepares them for those higher abstractions and mental efforts, of which they are at first incapable.

Most of the errors and conflicts in the Schools of Philosophy have arisen from the double or incomplete sense in which words are employed. The terms are all defined in a common language, but there is no fixed standard beyond the language itself. Each term is viewed from a different stand point, and, like the rainbow painted on the clouds, is different to every spectator, though apparently the same.

Mathematics is free from all such sources of mistake and error. There is no other subject of knowledge in which there is that exact equivalency between the thought and its sign. Number and Space, in all their elementary combinations, may be presented to the mind by pictorial representations. The senses are thus brought to the aid of the conceptive powers, and by means of this double language, the forms, attributes and laws of magnitude, are explained and verified.

The study of mathematics accustoms us to the strict use of this exact and copious language, in which all the terms are exponents of distinct crystallized ideas. Using these terms as instruments of reasoning, we advance with a steady step, secured from the sources or causes of error which are concealed under uncertain or conflicting meanings.

Knowledge is a clear and certain conception of that which is true. Its elements are acquired through the medium of the senses, by observation, experiment and experience; and these three indicate certain relations which the elements bear to each other, and which we express under the general name of law. Law, therefore, is a term of generalization, denoting an order of sequence in phenomena, whether in the material or spiritual, the animate or inanimate world. This order and connection are not obvious to the senses. They are the hidden treasures of knowledge, and are only discovered and brought to light by the highest exercise of the reasoning faculty.

Since the time of Aristotle, the exact law which governs the reasoning faculty has been well known. By careful analysis and a profound generalization, he subjected every principle of deductive reasoning to a single law, expressed by the dictum, and indicated every operation of that law in the syllogism. The system was yet incomplete. The major premise, on which the whole fabric rested, was assumed, not proved. Bacon supplied this deficiency, in showing that all our knowledge rests, ultimately, on the hypothesis of the uniform operation of the laws of nature, and that such uniformity may be inferred by the reasoning faculty, from a collection and comparison of facts, furnished by observation, experiment and experience. This completed the golden circle of logic, and subjected all the laws of nature to the processes of science.

It becomes, therefore, an important inquiry how far the study of mathematics is a means, in the cultivation of the reasoning faculties, through which we derive our scientific knowledge—how far it is a useful gymnastic of the mind—what mental habits it inculcates, and what developments it produces. We have already adverted to its clear, precise, and comprehensive language, and to the elementary ideas, which that language impresses on the mind. Are these ideas isolated—incapable of classification and wanting in the attributes necessary to a logical arrangement?

It is the chief excellence of mathematical science, regarded as a means of mental training, that the definitions and axioms are the prolific sources of every deduction. They are the ultimate premises to which every principle can be referred, and the law of connection which binds together all the truths of this complex system, is the simple law of the syllogism.

Mathematical reasoning, so far as the logic is concerned, is precisely the same as any other kind of reasoning. It differs from other methods only in the greater preciseness of its language, the nature of the subject and the more obvious relations of the premises to each other, and to the conclusion. It has been urged that these differences are detrimental, rather than useful, in the development of the reasoning faculty—that the exact equivalency between the idea and the language, the fixed and obvious relation of the premises to each other and to the conclusion, leave no scope for originality in the mental processes, and that truth is thus evolved mechanically, rather than intellectually. Another objection has also been found, in the fact that the matter in the mathematical processes is certain, while in all other cases it is contingent—and that to deal with what is certain, in accordance with obvious and fixed laws, disqualifies the mind to deal with what is probable according to laws less obvious and rigorous.

In regard to the second objection, it is quite certain that the degree of probability, in any given case, can only be determined by comparing what is contingent with what is certain—certainty being assumed as the standard—all inferences are relied upon as they approach this standard, and distrusted as they recede from it. Hence, in all systems of intellectual training, having in view the cultivation of the reasoning faculty, the mind should be accustomed to contemplate that which is certain, in order that it may form a true estimate of that which is contingent or probable.

How far the laws which regulate and control the processes of mathematical reasoning are merely mechanical, and how far their study and contemplation confine the mind to a mere routine, is best answered by a careful and searçhing analysis. The processes begin with obvious and elementary truths, defined by a precise language, and aided, if need be, by pictorial representations. They then advance step by step in a series of regular and dependent gradations, developing the concealed and sublime properties of number and space. These trains of demonstrative reasoning produce the most certain knowledge of which the mind is capable. They establish truth so clearly that none can deny or doubt. The premises are not only certain, but the most obvious truths which can be presented to the mind, and the conclusions result from the most palpable relations of the premises to each other. What discipline can better train the mind to diligence in study—to close and continuous attention—to habits of abstraction—and to a true logical development?

A wide distinction must be made between those processes of mathematics which are merely mechanical and that knowledge of the laws of the science which develops and applies those processes. The calculating machine is a mere instrument, but the discovery and application of the laws of its construction are among the highest efforts of genius. If the machine were dashed to pieces, it could be remodeled, for the law of its construction is known. The conception, therefore, is not mechanical because it is manifested by mechanical agencies. Descartes brought all space within the range and power of analysis, by new methods of representing lines and surfaces. Newton's sublime conception of the law of universal gravitation is developed in the language of Geometry. Does it follow, because the processes of Geometry and the rules for solving equations are reduced to fixed principles and settled methods, that the ubjects to which they may be applied are limited in their nature? or, that the contemplation of these subjects, through this, the only language in which they can be presented to the mind, is likely to give a contracted or one-sided development?

Mathematical Science deals with Number, Space, Time and Motion. Each is a type of the Creator, infinite in itself, and all are under the dominion of universal laws. In the development of these laws, in a language free from obscurity, and in a logic above the influence of passion, sophistry and prejudice, the mind acquires an intensity and ardor which lift it above the strife and petty controversies of earth, into the sphere of the intellectual and absolute. A theorem demonstrated is an indestructible truth; but this is not all, it is connected with antecedent truths of the same kind, and is also a guaranty of our success in new efforts to enlarge the boundaries of knowledge.

In the construction of the mathematical science, we begin with the axiom and proceed from proposition to proposition, under the guidance of a rigorous logic, till we reach the boundaries of that intellectual region which has been already explored. Here we pause, but do not stop; for beyond are hidden truths which excite our innate desire to know, and an ambition and hope of progress. So, when we stretch out the mathematics to explain and embrace the philosophy of the heavens, we proceed from our own planet, in regular gradations, till we reach the remotest orb of our system. Still further on, we enter the region of Arcturus, Orion, the Pleiades and the Milky Way; and, even beyond the smallest star whose light has reached the earth, is unmeasured space, yet perhaps to be surveyed by more perfect instruments, and measured by the known laws of mathematical science.

"There is good room to ask whether the peculiar energy of what might be called the mathematical soul does not carry with it a deep meaning, and declare the truth of man's destination at the first, and of his destiny still to take a place and to act a part in a world of manifested truth and eternal order. Do we venture too far in saying that, when mathematical abstractions of the higher sort take possession of a vigorous reason, there is placed before us a tacit recognition (one among several, all carrying the same meaning,) of the fact that the human mind is so framed as to find its home nowhere but in a sphere within which the absolute and the unchangeable shall stand revealed in the view of the finite intelligence?"*

The term "practical," in its common acceptation, often denotes shorter methods of obtaining results than are indicated by science. It implies a substitution of natural sagacity and "mother wit" for the results of hard study and laborious effort. It implies the use of knowledge before it is acquired—the substitution of the results of mere experiment for the deductions of science, and the placing of empiricism above philosophy. But give to "practical" its true and right signification, and it becomes a word of real

* Isaac Taylor.

import and definite value. In its right sense, it denotes the best means of making the true ideal the actual: that is, of applying the principles of science in all the practical business of life, and of bodying forth in material form the conceptions of taste and genius.

Beyond the obvious application of simple and known principles, the whole problem of the practical lies in the measurement, modification and best uses of the forces of nature. In all the uses and applications of these forces, material substances are employed, and these must be fashioned according to certain forms indicated by scientific formulas. These formulas are constructed from the laws which regulate the cohesion of the particles of the substance employed—the nature of the force to be applied—the amount of that force and the ultimate end to be attained. All these fixed laws of force—all their combinations—and all the forms of the materials employed in using them for practical purposes, can only be reached through the processes and language of mathematics.

Machines and workshops afford marked illustrations of the utility and value of mathematical science, and, in their resolution of difficult practical problems, furnish a striking exhibition of the power of mind over matter. Any one, introduced for the first time to the interior of one of our great factories, would doubtless regard with no small perplexity the equipment and play of so many variously directed instruments of motion—the great size and extent of the whole structure—the jar which startles at first, but by the steadiness of its pulsations soon persuades you to take the cadence and measure of the great machine, and to appropriate, as it were, a share of the producing power—and it would be strange if you were not also persuaded that all this bewildering procession of complex returning movements must be under the guidance of some great scientific law.

All the parts of that complicated machinery are adjusted to each other, and were indeed so arranged, according to a given plan, before a single wheel was formed by the hand of the forger. The power necessary to do the entire work was first carefully calculated, and then distributed throughout the ramifications of the machinery. Each part was so arranged as to fulfill its office. Every circumference and band and cog, has its specific duty assigned to it. They are connecting parts of an entire practical scientific system, over which one of the parts, fitly called the governor, is most ingeniously appointed to preside. It is the function of this apt and beautiful contrivance to regulate the force which shall drive the whole, according to a uniform speed; and it performs the office with such sensibility and seeming intelligence, that, on the slightest increase of velocity, it commences and executes, with easy gradations, a diminution of the moving force of the machine, and as instinctively calls up additional power the moment that the speed slackens. All this is the result of calculation. When the curious shall visit these exhibitions of ingenuity and skill, let them not suppose that they are the offspring of chance and experiment. They are the embodiment, by intelligent labor, of the results of the most difficult investigations of science.

The Steamship affords another impressive illustration of theoretical and practical science. Observe her form—how perfect in all its parts—how beautiful in outline—how exact in proportion. See how gracefully she rests upon the water, which she scarcely seems to touch. On the upper deck, the masts and ropes, the yards, the spars, the booms and sails, are all adjusted to the proper angle and are the instruments by which the power of the wind is pressed into the service of commerce. But this is not the power on which she relies. The great mechanical contrivance, to which I have alluded, which just now shook the earth with its jar, is to be readjusted and folded within a structure having its own peculiar form and limits, designed for special functions and moving on a new element. The source of power is a simple change in the form of a fluid.

The massive cylinders, the huge levers, the lifting and closing valves are contrivances to convey this power to the water wheels, where the resistance of the water, according to known laws, transfers it to the ship itself.

Over all this complication of machinery—over all this variety of principle and workmanship, science has waved her magic wand. There is not a cylinder whose dimensions were not measured—not a lever whose power was not calculated, nor a valve which does not open and shut at the appointed moment. There is not, in all this structure, a bolt, a screw, or rod which was not provided for before the great shaft was forged, and which does not bear to that shaft a proper proportion.

The language of Geometry and Number furnished the architect with all the signs and instruments of thought necessary to a perfect ideal of his work, before he took the first step in its execution. It also enabled him, by drawings and figures, so to direct the hand of labor as to form the actual after its pattern—the ideal. The various parts may be constructed by different mechanics, at different places, but the law of science is so certain that every part will have its right dimensions, and when all are put together they form a perfect whole.

When the work is done and the ship takes her

departure for another continent, a small piece of iron, a few inches in length, poised on its centre, under the influence of a known force, is the little pilot which guides her over trackless waters. Science has also provided, for daily use, maps and charts of the port which she leaves, of the ocean to be traversed and of the coasts and harbors which are to be visited. On these are marked the results of much careful labor. The shoals, the channels, the points of danger and the places of security, are all indicated. Near by hangs the Barometer, constructed from mathematical formulas, to indicate changes in the weight of the atmosphere and give warning of the approaching tempest. In close proximity are the Sextant and the Tables of Bowditch. These are the simple contrivances which science has furnished to correct the errors of the needle, by observations on the heavenly bodies, and to determine the exact position of the vessel at any moment of the voyage. Thus, practical science, which determined the form of the vessel best adapted to a given velocity, which measured and distributed the propelling force and which guided the hand of the mechanic in every workshop, is, under Providence, the means of conducting her in safety over the ocean. It is, indeed, the cloud by day and the pillar of fire by night.

The construction of railways is a recent and most important application of science. The mechanic arts, commerce and civilization have all received an impulse in this new development of power. The chariots of commerce, which rush with such dizzying velocity over the iron bands which now nearly encircle the globe, are all guided by immutable laws that have been carefully developed by the aid of diagrams and equations. When you see the long train, with its locomotive, ascending the mountain, fear not, for science traced the curve and balanced the forces. When the mountain is to be pierced instead of being scaled, a few lines drawn on paper indicate the precise points, at the opposite extremities, where the work is to be begun; and after years of labor the two working parties meet near the centre, and in the exact line established before the ground was broken.

In every case where power is employed, either to produce motion or to maintain a state of rest, the mechanical principle of force and resistance must be considered and discussed. Mathematics is the only form of language which connects science with all the mechanic arts and guides the hand of labor as it bodies forth the conceptions of the mind. It is, therefore, the only true basis of the practical; and perhaps it is not too much to add, that whatever is true and just in the practical is the actual of an antecedent ideal.

Material objects are the first things which attract our notice. We behold the earth filled with products and teeming with life. We note the return of day and night at regular intervals—the coming of summer and winter, and the succession of heat and cold. We see the sun in the firmament—we turn our eyes to the starry heavens and behold the sentinels of night as they look down upon us. Facts, often observed, suggest the idea of causes—and, when science scatters her light over the pathway of the past and the future, we learn the existence of general laws imparted by the fiat of Him who created all things—and come to understand that mind in all its attributes, and matter in all its forms, are subject to those laws—and that their study is the noblest employment of our intellectual nature.

To the uneducated man, all the world is a mystery. He does not see how so great a uniformity can exist with the infinite variety which pervades every department of nature, animate and inanimate. In the animal kingdom no two of a species are exactly alike; and yet the general resemblance and conformity are so close that the Naturalist, from the examination of a single bone, finds no difficulty in determining the species, size and structure of the animal. So, also, in the vegetable and mineral kingdoms, where all the structures of growth and formation, though infinitely varied, are yet conformable to like general laws.

The wonderful mechanism displayed in the structure of animals was but imperfectly understood, until analyzed and illustrated by the principles of science. Then, a general law, applicable to every case involving power and motion, was found to pervade the whole. Every bone is proved to be of that length and diameter best adapted to its use—every muscle is inserted at the right point, and works about the right centre—the feathers of every bird are shaped in the best form, and the curves in which they cleave the air are the best adapted to velocity. It is demonstrated, that in every case, and in all the varieties of form, in which forces are applied, either to increase power or gain velocity, general laws have been established to produce the desired results. Thus science makes known to us the foreknowledge and wisdom of the Creator.

But inanimate nature also speaks to us in the language of general laws, and it is in the investigation and interpretation of these laws that mathematical science finds its widest range and its most striking applications. Experience, aided by observation and enlightened by experiment, is the recognized fountain of all knowledge of nature. On this foundation Bacon rested his philosophy. He saw that the deductive process of Aristotle, in which the conclusion does not reach beyond the premises, was not progressive. It might, indeed, improve the reasoning process, cultivate habits of nice discrimination and give great proficiency in verbal dialectics; but the basis was too narrow for that expansive philosophy which was to unfold and harmonize all the laws of nature. Hence, he suggested a careful examination of nature in every department, and thus laid the foundations of a new philosophy. Nature was to be interrogated by experiment; observation was to note the results and gather the facts into the store-house of knowledge. Facts, so obtained, were subjected to analysis and collation, and from such classification general laws were inferred, by a reasoning process called Induction.

This new philosophy gave a startling impulse to the mind, and to knowledge. Its subject was nature —material and immaterial; its object, the discovery and analysis of those general laws which pervade, regulate and impart uniformity to all things; its processes, experience, experiment and observation for the ascertainment of facts, analysis and comparison for their classification, and the reasoning process for the establishment of general laws. But the work would have been incomplete without the aid of deductive Science. General laws, deduced from many separate cases, by induction, needed additional proof; for they might have been inferred from resemblances too slight, or from coincidences too few. Mathematics affords such proofs.

Every branch of natural philosophy was originally experimental; each generalization rested on a special induction, and was derived from its own distinct set of observations and experiments. From being sciences of pure experiment, or sciences in which the reasonings consist of no more than one step, and that step an induction, all these sciences have become, to some extent, and some of them in nearly their whole extent, sciences of pure reasoning: thus, multitudes of truths, already known by induction, from as many different sets of experiments, have come to be exhibited as deductions, or corollaries from inductive propositions of a simple and more universal character. Thus, Mechanics, Acoustics, Optics and Chemistry, have successively been rendered mathematical: and Astronomy was brought by Newton within the laws of general mechanics.

The substitution of this circuitous mode of proceeding, for a process apparently much easier and more natural, is held, and justly too, to be the greatest triumph in the investigation of nature. But it is necessary to remark that although, by this progressive transformation, all sciences tend to become more and more deductive, they are not, therefore, the less inductive: for every step in the deduction rests on antecedent induction.*

We can now, therefore, perceive what is the generic distinction between sciences which can be made deductive, and those which must, as yet, remain experimental. The difference consists in our having been able, in the first case, and not in the second, to establish a set of first inductions, from which, as from a general law, we are able to draw a series of connected and dependent truths. For example, when Newton, by observing and comparing the motions of several of the bodies of the solar system, discovered that each, whether its motions were regular or apparently anomalous, conformed to the law of moving around a common centre, urged by a centripetal force, varying directly as the mass and inversely as the square of the distance, he inferred the existence of the law for all bodies; and then demonstrated, by the aid of mathematics, that no other law could produce such motions. This is the most striking example which has yet occurred of the transformation, at a single stroke, of a science, which was in

* Mill's Logic.

a great degree experimental, into one purely deductive.

It is in the great problem of the solar system that mathematical science displays its omnipotent power. The sun himself, manifesting his inexpressible glory by the floods of golden light which he scatters through the immensity of space, is yet subjected to the analytical formula, and must confess to it, from his more than imperial throne—his exact dimensions—his weight and balancing power, and his relative importance when compared with the smallest mote which his own light has revealed. It is thus that the intellectual power, aided and stimulated by the processes of mathematical science, has been able to trace backwards, to the earliest past, all the motions of the heavenly bodies and to bring the remotest future of the planetary system within the range of its computations. It is thus that man, inhabiting one of the smallest planets of the system, computes the celestial cycles and determines all the laws of the movement of the celestial machinery.

He has done even more than this. Those vagrant bodies of the heavens which occasionally visit our system, and which seem to have escaped from their own spheres and to wander heedlessly through space, are yet subjected to the power of analysis. A few observations, made by the practical astronomer, afford the necessary elements for computing the forms of their orbits and their periodic times; and in distant years, at the indicated moment, the comet again blazes in the sky. In short, before this august power all nature yields up the mystery of her laws. If, then, we would enter her spacious temple, and seek after the knowledge which is there, let us not forget the Aladdin's lamp of mathematical science, which, being properly touched, will disclose more treasures than have ever been described in Eastern fable.

The place which mathematics should occupy in a system of collegiate instruction is an inquiry of the gravest import, and necessarily involves the question, What should be the nature of the system itself?

It was stated, in the opening address, on the highest authority, "that the end of a liberal education is the general and harmonious evolution of all the faculties and capacities of the mind in their relative subordination." It is not the base, nor the massive shaft, nor the beautiful forms of the capital, which fill the mind as we gaze on the Corinthian column; but it is their unity and the general effect of their combination. It is the whole mind, in all its intellectual and emotional faculties, to which the experienced educator addresses himself.

So far as our knowledge extends, we have found

in that mysterious essence, the mind, a faculty adapted to the apprehension of every law, and an emotion corresponding to the contemplation of every object. May not the reverse of this proposition be true? May it not be, that for every faculty of the mind, whether intellectual or emotional, there exists, somewhere, a proper object of contemplation? and that the perfection of our knowledge and being will be attained when all such objects are found? It is in accordance with this law that different studies cultivate different powers of the mind, and that it requires the study of many subjects to give a general and harmonious evolution of all its faculties. Mathematics does not equally cultivate every faculty—it is the massive trunk and outward form, but language, literature, and moral culture, are the sap which ascends within, and which is necessary to give beauty to the foliage and health and harmony to the whole development. All the colors of the rainbow, which are painted on the clouds, are necessary to the perfect light of day—so every light of knowledge is required in the perfect illumination of the mind.

It is the special function of mathematical studies, to cultivate the faculty of abstraction and the habit of intense and continued attention—to establish in the mind a self-centering power that shall subordinate all the intellectual faculties to the control of the will

—to create, as it were, a governor of the intellectual machinery, that will give harmony and uniformity to all its motions. As an elementary formula of logic, it is the most simple and perfect. As a drill, in the structure and use of language, in its primary forms, no exercise insures greater precision in the use of words, or imparts to the mind as certain relations between the signs and the things signified. In its higher branches, it is even an aid in the study of theology ; for it constantly raises the mind to the contemplation of the Unchangeable and the Infinite. Mathematics, therefore, is an aid and auxiliary in every other branch of study. It may be pursued too exclusively—the mind may become too much absorbed by its machinery and formulas ; but this danger is common to the study of every other subject. A life spent exclusively on the Greek Grammar would not make a Greek scholar ; nor can the wide field of deductive reasoning be explored by repeating the formulas of the dictum and syllogism.

Concurring fully in what was said, in the opening address, concerning the great value of the study of the Greek and Latin languages, and also in the merited eulogium of the manner in which these languages are taught in this institution, I may yet be permitted to say, that there is another language far more comprehensive than either or both of them :

the language of mathematics, which embraces within its ample folds all the laws of the material universe. This language takes us back to the birth of matter, and measures and records every step which each planet has taken since it began to move. Yea, more: it is prophetic—it reveals all future motions, and indicates the precise places which all matter must occupy, at any given instant of future time.

This is the language in which the practical astronomer studies the heavens. It is the telegraphic wire which has enabled him to communicate with every planet of our system—to measure its diameter, its specific gravity, the dimensions of its orbit, its times of revolution and its balancing power in the system of the universe. It is this language which has enabled him to bring the ring of Saturn into his own study, where he sees it face to face, and, as it were, touches the very particles of matter of which it is composed.

This language has enabled the naturalist to trace the dominion of law over all matter endowed with life. The contemplation of the minute objects of creation may appear, at first sight, unworthy the labors of the highest genius—but it is quite otherwise. The turtle's egg, the little gnat whose tiny wings vibrate five hundred times in a second, and the entire solar system, are each an embodiment of a thought of God. Whether we look through the microscope or the telescope, we are equally instructed in the wonders of creative power and universal law.

But science is not all in all. It does not compass the final aim and ultimate end of our being. Though it reaches back to the time when God said "Let there be light and there was light," and forward to the time when "there shall be a new heaven and a new earth"—though it measures all space—though it explains all laws relating to matter and motion—though it transports us to the central point of the physical universe, whence we behold the heavenly hosts moving in celestial harmony: yet, when we approach that mysterious line where the finite terminates and the infinite begins, new visions open to the mind—all science and human knowledge fade away like castellated clouds made brilliant by the setting sun—Faith then arises in supernal beauty, and, with veiled eyes and trembling voice, we confess, "In the beginning was the Word, and the Word was with God, and the Word was God."

INAUGURAL DISCOURSE

BY

CHARLES MURRAY NAIRNE, M. A.,

PROFESSOR OF LITERATURE AND PHILOSOPHY,

February, 1858.

ADDRESS.

The subjects assigned to the Professor of Literature and Philosophy in Columbia College are so multifarious, that a notice of each in succession, no longer than a brief newspaper article, would occupy the whole time allotted to the present discourse. Within the narrow bounds to which I am necessarily limited, how much could a man say on the great topics of Universal Grammar, Rhetoric, Logic, Oratory, Æsthetics, Psychology, Ethics, Ancient and Modern Literature, the History of Philosophy, and that finest of all the Fine Arts—the living representation of our thoughts and feelings, by the symbols and the music of our mother tongue? I know no process of intellectual condensation, by which any adequate or interesting account of so many departments of learning could be laid before you, on such an occasion as this. Every one of them would, of itself, furnish materials sufficient to fill full that utmost limit of the American listener's patience—an hour.

Were I to make a selection from the encyclopædia

of arts and sciences that I am appointed to teach, and to group together those three kindred branches, Logic, Grammar and Rhetoric, however unpopular my choice, from so great a variety, might seem, I doubt not that I could unfold to you such views of the human mind, in its operations of thought and expression, as would not fail to excite your curiosity, and command your attention. It would be my duty to show you that man is distinguished from the lower animals, and connected with the nature of angels and of God, by the reasoning faculty; and that, in the use of this faculty, all mankind—from the child to the sage, from the barbarian to the philosopher—are doing precisely the same thing in the self-same way—namely, deducing conclusions from premises. The learned are conscious of the syllogistic process, and can reduce their reasonings to the syllogistic form; the unlearned are unconscious of the process, and perform it naturally; nevertheless the process is identical in both; and its discovery is one of the noblest examples of generalization within the whole compass of human knowledge—quite as noble as the law of affinity, which, in fixed proportions, holds together the constituents of matter, or the law of gravitation, which links, by an invisible bond, the spheres of the celestial concave.

I should further have to show you that, as language

is the body of thought—the audible or visible symbolization of the unseen spirit's operations and states—and that as the process of thinking is human—common to, and characteristic of, the entire family of man—the propositions, in which thoughts are embodied, must have the same essential form, and consist of the same elements, in every language under heaven. It would thus appear that Grammar is not an art but a science—a department of the great science of mind, possessing deep interest as an intellectual study; and not merely a system of rules for the exercise of school-boys, and the prevention of slips of the tongue. It would appear that, while various nations employ various sounds to designate objects, actions, attributes, and relations, and have thus each a different lexicography, the grammar, properly so called, of all languages is, with the exception of a few idioms and peculiar arrangements in each, the very same; and that it is, in fact, nearly as absurd to talk of Greek Grammar and Latin Grammar, English Grammar and French Grammar, as it would be to talk of Greek, Latin, English and French Logic—Greek, Latin, English and French Chemistry. Logic is logic, and Chemistry, chemistry, in whatever tongue they are employed or expounded; and so, also, Grammar is grammar—the science of the human speech—in Latin or in English, in Greek or in French, in Chinese or in Choctaw. Wherever men speak—wherever the μέροπες ἄνθρωποι exist—they must of necessity indicate objects, and, therefore, have nouns; actions, and, therefore, have verbs; attributes, and, therefore, have adjectives; relations, and, therefore, have prefixes and suffixes separate or conjoined; and the subject, predicate and copula must be used, as often as mankind have anything to speak of, and something to say concerning it. The distinctions of gender, number and comparison—of person, time, mode, and voice are not arbitrary, but determined by the nature of things. In short, the principles of grammatical science are universal and necessary; and when the grammars of various languages are divested of the absurdities with which pedantry has overlaid them, it will be found that the difference between one tongue and another is simply a difference of vocabulary and arrangement—something to be mastered by the memory, rather than grasped by the understanding—something that can not be reduced to law, unless we receive as philosophy the hypothesis that certain vocal elements are the natural and universal representatives of certain ideas.

I should still further have to show you that, as reasoning and speech are essential attributes of humanity, so, in the use of these for the purposes of convincing and persuading, the same methods of inventing arguments, and the same ways of arranging and applying them, are common to every speaker under the sun—to all nations, and kindreds and peoples and tongues—to the Indian chief who harangues his tribe, the diplomatist who negotiates treaties, the legislator who evokes the applause of senates, and the minister of religion who commends salvation to dying men. Real Rhetoric is no conventional mode of dressing up Truth—no mere fashion, changing from year to year, and varying capriciously from beauty to deformity; but a genuine, legitimate Art, founded on universal and immutable principles. It is an art, indeed, to which genius sometimes may attain almost spontaneously, as Homer and Shakespeare did in poetry; nevertheless, like poetry, it has its conditioning laws which the philosopher investigates with pleasure, and which even genius may study with advantage. For genius is no lawless, wayward power. Its own insight discerns the ideals of truth and beauty, and these it publishes to mankind in its own practice. It is the image and vicegerent of Eternal Wisdom, proclaiming the law of Heaven to others, while itself yielding to it a free and loyal obedience.

Or again, quitting abstruse discussion, and choosing a more attractive flower from my garland, I might entertain you with the history and principles of Oratory. Were I to make this selection, it would be my task to describe those mighty masters of eloquence to whose fervid speech the hearts of men have thrilled, and by whom a power was wielded to shape the destinies of nations and the world. I should tell you of Nestor and Ulysses as they utter melodious fascination in the verse of Homer; of Demosthenes, who

> "fulmined over Greece
> To Macedon and Artaxerxes' throne;"

of Tully, who transformed Athenian vehemence and splendor into Roman stateliness and majesty; of One far greater still, who, sitting on the mountain side, or by the crowded shore, proclaimed as man never spake, and with a celestial dignity beyond the loftiest repose of art, the sublime revelations of life and immortality; and hastening down the stream of time I should glance, as I passed, at the famous preachers and disputants—the Augustines, and Chrysostoms, and Abelards—of the middle ages, till, having crossed the abyss that divides the ancient world from the modern, I should group before you, in their various characters, the most distinguished orators who have flourished since the birth of the Reformation—Luther and Knox, with their rugged impetuosity; the more courtly and classic rhetoricians of the Anglican and

Gallican churches, and the stern conscience-searchers of the Puritan meeting-house; the fiery invective of Chatham, and the magnificence of his indomitable son; the glory of Fox, the splendor of Sheridan, and the philosophic gorgeousness of Burke; the forensic brilliancy of Erskine, Curran, and Scarlett; the energetic elegance of Canning, and the dark strength of Brougham; the fearless simplicity of Henry, the logical massiveness of Webster, the prophetic rapture of Edward Irving, and the overwhelming intensity of Chalmers.

And when I passed from the distinguishing characteristics of these and other great masters in oratory, to the nature of eloquence itself, I should show you that the grand secret of power in them all was naturalness and earnestness; and that the attributes which peculiarly belonged to the worthiest of them, were resolute honesty, strong love of man, and a heart-felt adoration of truth.

Or again, if, omitting the laws of reason and speech, and the practical use of these laws by the orator in convincing and persuading his fellows, I were to select from my repertory the subject of Æsthetics, or the Philosophy of Taste, it would be my endeavor to display before you that beauty which clothes all Nature as with a vesture of light, and has its source and centre in the Eternal, who dwells amid light that is inaccessible and full of glory; and to investigate that susceptibility, unpossessed by the brute—whose eye conveys no sense of loveliness from the loveliest landscape—but bestowed on human beings, and regaling their souls with all those delights of shape and sound, of motion and melody, which reflect, in Creation, the ineffable aspect of the Infinitely Beautiful. Nor would my essay be complete, till, in addition to the objective beauty of God and His works, and the subjective human sensibility that thrills to it, I should speak of those immortal creations, wherein the genius of poet, painter, sculptor, architect, musician, and orator, has enshrined the divine loveliness and sublimity of the universe; and show you how the spirit of every one of them, either consciously or unconsciously, held high converse with Him who, from the beauty of His holiness, sheds over heaven a brightness above the brightness of the sun.

There is what may be termed a language of form, expressed in figure and tone, addressing itself intelligibly to the reason, and exciting in the heart emotions corresponding to every sentiment of rational beings. Man, as rational, has the capacity to understand this language, and, therefore, it is, to a certain extent, known and read of all men; but the language of form must be studied in order to be fully comprehended, and the susceptibility must be cultivated, in order to receive all the enjoyment which the language is fitted to awaken. From the intercourse that we are compelled to hold with our fellow-mortals, we learn first to interpret the symbols of beauty in the lineaments of the human countenance, and in the accents of the human tongue. That mysterious thing which we call expression is evidently conveyed by mere shape and sound; and, to become sensible of the wonderful adaptation of these to represent every shade of sentiment, we have only to consider how slight are the modifications of outline which will alter the whole expression of one's face, and the changes of tone which will represent joy or sorrow, cheerfulness or solemnity, hope or despondency. It is the same countenance that we see, and the same voice that we hear—the countenance and the voice of our familiar friend—not a feature or tone is unrecognized; but complicated changes of form have taken place, which the reason instantaneously apprehends, and to which the susceptibility instantaneously responds. The changes in point of quantity have been very small, but they have been sufficient to tell the story of one mind to another; and to tell it with a rapidity and concentration to which the power of ordinary language is but feebleness. Now, from this one example we may learn the general expressiveness of shape and sound; and understand how the Divine Artist, in creation, or the human artist, in his chosen walk of painting, sculpture, music, or poetry, may convey to all rational beings, by outline and measure, the ideal that exists in his own soul. In the course of our Æsthetical Education, the language of beauty becomes continually more pregnant to our intellect and more striking to our sensibility, till, at last, in the galleries of art, in the cathedral and the concert-room, or amid the scenery and harmonies of nature, the sentient spirit drinks in meaning and delight from all that surrounds it. The insight of reason reads the sentiment of every form. The statue, the picture, the tune, the landscape, are all inspired—and the mind catches the import of each "peculiarity of modulated tone and delineated figure. The utterance of human sentiment in sensible forms gives Beauty; and when the disclosed sentiment is that of a superhuman spirit, and we stand awe-struck in the presence of an angel or a divinity, the Beauty rises proportionally, and elevates itself into the Sublime."*

Or yet again, were I attracted from all the rest of my themes by the charms of Literature, it would be my duty to characterize, in the first

* Hickok's Pyschology.

place, the collective literature of nations, as embodying and exhibiting the peculiarities of national mind:—the primitive simplicity of the Hebrew chroniclers, and the unapproachable majesty of the Hebrew poets—the splendor, variety, and all but perfect beauty of Grecian genius, and the borrowed lustre of its Roman imitators—the half-christian, half-pagan imaginations of mediæval Italy and Spain—the grandeur of English letters in the early vigor of their youth, when Shakespeare created, and Bacon philosophized, and Raleigh began the history of the world—the more artificially polished productions of the Gallic muse, which, crossing the channel as the missionaries of a less sturdy civilization, converted the English Miltons and Jeremy Taylors into Popes and Addisons, and the Scottish Knoxes and Buchanans into Robertsons and Blairs—the Teutonic revulsion, which brought back the reign of originality and of power in Germany, and spread from thence to Britain and even into France herself—and last of all, that hybrid style of thought and writing, which the mixed population and rapid growth of our own country have necessitated, and the elements of which have not yet become so blended and assimilated into a unity as to constitute a peculiar national literature. And then, passing from the broader distinctions of national genius to the more marked peculiarities of individual authors, it would be my happiness to expatiate in retrospect among the "departed spirits of the mighty dead"—of all whose names live in the page of history, and without whom History herself had never been—seeing that if the exploits of kings and heroes had remained unchronicled by annalist and bard, they would all have been forgotton utterly, or only recalled, in dimness and in terror, by the ruins of ancient cities,

> "And mighty relics of gigantic bones,"

turned up by the peasant's plough from the battle-fields and burial-grounds of unrecorded generations.

> "Vixere fortes ante Agamemnona
> Multi: sed omnes illacrimabiles
> Urguentur ignotique longa
> Nocte, carent quia vate sacro."

Having thus dismissed, by little more than a mere mention of them, all the other topics belonging to my department, I come now to the noblest and most arduous of the whole—the philosophy of the True and the Good. Were I to attempt presenting you with an outline of intellectual and ethical science, and a skeleton history of philosophy, extending from the remote era of Pythagoras, who first employed the term, through the various schools of Greece and Italy, down to the present time, the sketch would be so meagre, imperfect and uninteresting, that it would give little satisfaction either to listeners or to speaker. Neither can I allow myself to enlarge, in general terms, on the importance of moral and metaphysical study, as dealing with the most momentous questions that can engage the mind of man, and investigating the foundations of all knowledge whatsoever. It is on the high places of philosophy that the skeptic, the atheist, the pantheist, the materialist, and the spiritualist must be met and overthrown. Hence the value and the difficulty of the inquiries which philosophy embraces. But, instead of touching, except incidentally, on any of these inquiries, I deem it of far greater consequence at present to aver, with as much publicity and emphasis as possible, that of all the professors in your College, it is most indispensable that he who occupies the chair of philosophy should be thoroughly sound in the faith of the Gospel; for, in the name of science and under covert of her robe, he may teach, if so minded, the wildest and most pernicious doctrines. I desire, then, first of all to declare that, both from the constitution of my intellect and the impulses of my heart, I am compelled to believe that there is a God—a self-conscious, personal, infinitely gracious Maker and Father of all. The infinite is not opposed to the finite as light is to darkness, truth to falsehood, right to wrong, virtue to vice. The infinite embraces the finite, and the idea of the latter necessarily calls forth, as its correlative and complement, the idea of the former. My mind cannot survey the boundaries of the finite and conditioned without gazing awfully into the infinite that contains it, and reverentially toward the Absolute, of whom I myself am a feeble image. My soul is not satisfied—its natural craving is not filled—until it has passed the confines which mark off that which is limited within that which is limitless. Here am I placed "upon this bank and shoal of time" between two eternities—the denizen of a little isle amid an immeasurable ocean! My vision can reach but a brief space into the vast profound that environs me above and below, on the right hand and on the left; and although my spirit, as it makes excursions into creation, can discern much that is good and great, fair and admirable, it is still perplexed and baffled in its contemplations—"shadows, clouds and darkness rest" upon its views, till, from its own depths, like the sun from the nether hemisphere, springs the sublime discovery that there is a God!

In the stillness of a star-lit night, you may have cast your eyes over some fine landscape, and as you traced the glimmering circuit of the woods, and recognized the dark masses of the mountain-range, and saw the stars reflected in the river's bosom, and descried the mansions, turreted and gray, or less picturesque and less hallowed by time, rising through the shade, and humanizing the whole scene with the interests and occupations of man—as you stood, gazing and musing, you have said within yourselves—"How fair would this prospect be were the round moon now pouring her lustre on river, and wood, and dwelling, and hill; and how passing fair, when it lies glowing in the full sunshine that at once discloses and exalts its loveliness!" Nay, the very pleasure with which, even in the night, you behold it, is mainly owing to your recollection of its daylight glories, or of something similar; and you can scarcely fancy the dim and dull impression it would make upon a being who could not fill up its proportions from such recollection, and body forth its hidden features, in the exercise of an imagination which had been informed by the actual survey of the unveiled beauties of nature. *It is even so with Creation when contemplated apart from a Creator! It is even so with the present condition of things when regarded apart from a God of justice and goodness, holiness and truth—the very God whom the Bible describes.* Without a God, there lowers a most perplexing obscurity over the whole. I can discern beauties, but they are clouded; harmonies, but, when I attempt to track them, they fade in the infinity of the surrounding darkness; design, but it is only fragmentary, and not seldom apparently frustrated; operations, benevolent, and, to some extent, effectual, but often cruelly interfered with, and rendered distractingly abortive; something grand and graceful, it is true, but shadowy and evanescent, dreamy and dubious, without beginning and without end; and I am puzzled to account for interruptions, and vacuities, and discrepancies, and disturbances, and feel intensely the need of some superior illumination to irradiate the entire field of view, and dispel the mystery—a mystery as much of confusion as of vastness—that broods over everything before me. Chains of causation I can partially trace, but I discern no Omnipotent Hand from which they are suspended; goodly fabrics of antecedent and consequent I can see, but no Rock of ages on which their foundations are laid; motion I perceive, but no Prime Mover; regularity, but no Regulator; law, but no Law-giver; life, but no Fountain of life; scattered portions of truth, but no great Being who is the substance of truth—in whom all truth centres, and of whose nature all truth is only the disclosure and the outward expression! *Now, the master-key to the whole of this mystery is the existence of a Supreme Creator and Ruler.* The forth-flashing of this grand fact is the dayspring from on high, which, illuminating the Kosmos, brings to our view its order and dependence—its origin and its end; enables us to walk surely, like those who walk at noon, instead of groping and peering like those who walk in darkness; and gives rest to the soul's weary wings, by presenting an ultimate object whereon, in common with the entire universe, the exploring spirit reposes from its travel, and is satisfied.

When I first look up to the heavens, I behold nothing save an expanse of splendid confusion—a high o'erarching canopy glittering with lights of spiritual brightness. Their distances are all the same to my vision, and they appear scattered over the mighty concave at random. No sound issues from the aerial dome—no living thing can be discerned walking amidst these lamps; and when they themselves are, at length, discovered to move, their march is tardy and without array; for they fall not into ranks, and some of them seem to wander even from their own circles. Amid the multiplicity of luminaries, there is still obscurity. The stars are still the stars of night. Whence are they, I ask, and what are they? What is their nature and what their use? Is the frame-work, in which they are inlaid, really a firmament—a substantial, resisting roof—and do they stud its surface merely to regale my eyes, and exercise my curious fancy? I cannot tell!

As yet I cannot tell: but let me grasp the torch of

science. The astronomer demonstrates that those lamps are orbs—probably worlds like our own; that they revolve in paths of geometric symmetry, although so vast that the whole vault overhead is too limited a scroll to exhibit such a portion of those paths as would determine their figures to our sight; and that, throughout all space, there prevails a law which governs the huge globes wherewith its amplitudes are filled, and, under this law, that which originally appears disorder is regularity, far more accurate and exquisite than that of the most ingenious and delicate of human contrivances. Now I begin to approach towards satisfaction. The firmament, I find, is not a solid crystalline canopy; neither is there any longer disorder among the starry train. My mind now cleaves the depths of space, and, to the glance of science, mechanism, stupendous both in magnitude and harmony, is disclosed in its mighty and mysterious recesses. But after all I am not yet content. My spirit pants with the majesty of its own discoveries. I am confounded by the very grandeur which has been evoked. Amidst an illimitable universe I stand awe-struck and baffled, as if, too daring in my curiosity, I had intruded, under guidance of a potent genius, into a region of sublimity where even he might fear to tread. Here it is, however, that the still small voice of my inmost reason is answered by the celestial oracle of Revelation; and the two, blending into harmony, proclaim—"God is, and God reigneth!"—Within the infinite domain where I had penetrated, they point me to a throne, and to a Sovereign seated thereon. The Almighty Maker and Mover is seen! My wonder now becomes adoration; my astonishment is exalted into reverence. The insecurity, the uncertainty, and the absence of cause, which oppressed my soul, are now gone. It no more falters amid unexplained marvels. It has risen to the summit of truth, and from that empyreal height it sees, like a seraph on the battlements of Heaven, the whole creation roll beneath it, without shock and without confusion! The light which the astronomer kindled was sufficient only to show the vastness of the prospect. Dimness and doubt still lay upon its illimitably receding depths. It was still the landscape without the sun. The God who said

> "Let Newton be,"

was still Himself to be revealed; and then, but not till then, all became really light; and the orbs of the sky were perceived to obey His voice, and their splendor seen to be an irradiation from the "co-eternal beam of the Eternal"—the "light which no man can approach"—

> "Bright effluence of bright essence uncreate!"

It is thus that the existence of a God forms the Key-stone of the entire structure of knowledge. His being is the grand truth, that, like the central sphere of our solar system, gathers all others around it, and harmonizes them all, and sheds light upon them all, and infuses life into them all; and he, that would shut out this truth from his investigations, seems to me scarcely so wise as the man who should make his own chamber his universe, and content himself with examining its paltry appointments by the glimmer of his own taper, while he jealously excluded every ray coming from the fair and illuminated world beyond its walls.

The Bible tells us that, before man was made, the earth was replenished with every green and every breathing thing. The garden was planted and watered, and it teemed with life and beauty. Streams sparkled in the sun, breezes whispered in the shade, fruits glowed upon the boughs, flowers enameled the sward and opened their fragrant bosoms to the day, birds warbled among the bowers of Eden, beasts sported on its glades, and all creation awaited the advent of creation's lord, whose immortal mind was capable of ruling it, and appreciating the proofs of wisdom, power, and goodness which, though existing in their own frames and functions, the creatures themselves were unable to comprehend. And surely it is no vain or improbable imagination to fancy the first man picturing to himself, how unfinished and unsatisfactory would have been the curious work before him, had he who was its crown and glory not been produced, and invested with dominion over it. We can still further fancy his procedure, as, in the exercise of his newly-awakened consciousness, he must have inquired into the secret of his own being—gazing for a while on external things, and then turning to his own body, perusing his own limbs, trying his own powers, and, when he found all so fitly and surprisingly made, questioning the creatures already formed, as if they, with thought and speech like his own, could tell him whence and what he was, and conjecturing, in the fullness of his doubt and wonder, what all the enchantment about him could mean, till, amid his delight and perplexity, he at length hears and knows the voice of God, and, bending with instinctive reverence before His presence, learns from the Divine utterance the mystery of his own existence and destiny, and the explanation of the manifold other existences that encompassed him on every side. Such an incident as this would come upon him with all the cheerfulness and certainty of light. His undefined desires it would both bring to shape and satisfy, and, like the discovery of any other great principle, it would reduce to order, and clearness, and unity, that which, without it, or something equivalent to it, would have forever remained to him a problem incapable of solution.

Now, this stroke of Milton's imagination, which I have adapted to my present purpose, is not produced as a fact, but as an illustration. It is most eminently natural. To be sure, there is none of us in circumstances similar to those of Adam with reference to the knowledge and theory of creation. The existence of a Creator and Supreme Ruler is part of our earliest and most familiar belief; and thus it is that we are under the necessity of making a strong effort to appreciate the sudden and self-evidencing power of a discovery like that which we suppose to have been made to him. Nevertheless, on making such an effort, the result will be powerfully felt, and we shall perceive that, in order to give unity, consistency, and intelligibility to the universe, both in its physical relations and in its moral aspects, we are compelled to admit the being of a God. It is the principle of affinity which gives unity to Chemistry—of gravitation which gives unity to Astronomy—of conscience which gives unity to Ethics—of propitiation which gives unity to Christianity—of life which gives unity to animals—of personality which gives unity to the human being;—and, in like manner, the universe is not felt to be One—seems not a Kosmos, but a stupendous puzzle, until reason starts, and Revelation confirms, that greatest of all truths, that God is, and that God reigneth—the Maker, Mover, and Father of all.

In the second place, I seize this public and appropriate opportunity of declaring that, from no superficial study of its evidences both historical and internal, I am steadfast in the belief that the Bible is the word of God—that the inspiration of the sacred writers is no mere theological name for the intuitions of human genius—that "thus saith the Lord" means literally and simply "thus saith the Lord"—and that, with the trifling exception of accidental mistakes common to all books that have been multiplied by transcription, the Holy Scriptures contain truth without mixture of error. I am fully aware that the Bible was not given to instruct men in science and philosophy, and that its language is the language of the people, not of sages and *savans.* I am further most fully aware that we are now in possession of a critical apparatus—a method of interpreting ancient writings—which implies not only a grammatical familiarity with their dialect, but likewise a historical familiarity with the speculative opinions and modes of thinking, common to the age and country in which the writers of them lived. And I am still further aware that the researches of travelers and antiquaries, and the labors of scientific men—astronomers, geographers, geologists, naturalists, metaphysicians, ethnologists, and even chemists—have cast light on many portions of Scripture, and enabled critics to improve the interpretation of them, so that apparent discrepancies between science and revelation have been reconciled, and those things which, at first, were difficulties, have actually become demonstrations. Of all these facts I am most fully aware; and, in view of them all, I affirm that were my investigations in philosophy to land me in a result that is clearly at variance with the well-ascertained import of the Divine Word, I would stop short instantly, assured that I was either wrong in my philosophical principles, or faulty in my logical deductions; and I would earnestly retrace my steps, and search diligently till I had found where my error lay. Others may call this timidity—or even bigotry—if they choose. I call it reverential caution; and I freely confess that I should neither have the foolhardiness to intrude anti-christian theories upon the undergraduates of a College, nor the dishonesty to retain a position, where I should be compelled to inculcate doctrines which I did not most firmly believe.

According to my view—which is also that of St. Paul—and, therefore, the correct one—the grand central idea of the Gospel is atonement by sacrifice. Now, it is certain that the great "mystery of godliness—God manifest in the flesh," is altogether beyond the ken of human philosophy. German rationalists and their disciples may tell me that every man is an incarnation of Divinity, and their words, when they so speak, may not be destitute of meaning; but of this I am very sure, that they do not mean what St. John says, when he announces that the "Word was made flesh and dwelt among us." This truth is a matter of pure revelation. Nevertheless, although our philosophy could never have solved the divine problem for which the Word became incarnate—how shall God be just, and yet the justifier of sinners?—philosophy assuredly does point us, with no uncertain indication, to the necessity of a Redeemer, and hints not obscurely that our Redeemer must be Almighty. A short demonstration of these facts will terminate the present address, and show, in a sufficiently intelligible way, how philosophical investigation may be applied to questions of the highest practical moment.

The knowledge and the power of man being both limited, he may not be able, in the first place, to form a perfect conception of an end which he desires to accomplish; and, in the second place, he may not have sufficient skill to devise and adapt the means whereby it may be accomplished perfectly. His contrivances may be faulty, either by excess or by defect. The material chosen may not be the most suitable, and it may be improperly distributed. There may be a superfluity of strength in one part, and a deficiency in another, and the application of his machinery may, and in fact generally does, admit of improvement. In short, his advances towards perfection are necessarily tentative and experimental. He does not produce it at once by intuition or instinct, as the bee constructs its cells and the bird its nest. And as it is with man's material contrivances, so also it is with his schemes of moral and intellectual mechanism. In government, in education, and in philanthropic enterprise, he proceeds by trial and error, and does not arrive at the best plan till after many a failure and many an alteration.

But God, on the contrary, being infinite in wisdom and infinite in power, knows at once the very end He would gain, and the very means that are requisite to gain it. This is an obvious deduction from the very notion of Godhead. And the truth, thus emanating from a source *a priori*, is exemplified in all the contrivances and arrangements of the universe. In God's works there is neither defect nor superfluity. The power employed is most precisely proportioned and adapted to the work that is to be done. If the whale, for instance, requires to dive to depths in the ocean where the pressure would be destructive to other creatures, it is made strong in proportion to that pressure. If the eagle must soar heavenward, its bones and quills are made light, and if it must battle with the storm, they are likewise made strong. If the *habitat* of a fish is the dark waters of the Mammoth Cave, the creature is unprovided with eyes, but in the feline family, which seek their prey in the night-time, the organ of vision is capable of extraordinary enlargement. The tribes of the sea have no fountain of tears wherewith to lubricate the eye-ball, because they need none; but the dwellers on the land are furnished with the necessary secretion. And so on, throughout all nature, there is nothing superfluous and nothing defective. In cases of human mechanism, where calculations, involving the profoundest mathematical principles, have been made to determine the exact medium between excess and defect, it has been found that the Creator had anticipated the solution of the difficulty. "During the latter part of the last century,"—says Edgar Allan Poe—"the question arose among mathematicians,—'to determine the best form that can be given to the sails of a wind-mill, according to their various distances from the revolving vanes, and likewise from the centres of the revolution.' This is an excessively complex problem; for it is, in other words, to find the best possible position at an infinity of varied distances, and at an infinity of points on the arm. There were a thousand futile attempts to answer the query, on the part of the most illustrious mathematicians; and when, at length, an undeniable solution was discovered, men found that the wings of a bird had given it with absolute precision ever since the first bird had traversed the air." The cells of the honey-comb afford another and more familiar example of the same law. They are so constructed as to give the utmost room that is compatible with the utmost stability and compactness. There is no loss of space and yet no diminution of strength.*

* As if to demonstrate the existence and rigid authority of this law in the most emphatic manner possible, we find it extended even to the region of the supernatural. The miracles of Scripture, although exceptional, as unusual exhibitions of Divine power, are not exceptional in respect of the law which we are now considering. First of all, no miracle is performed unless the occasion plainly justifies and demands it. The rule that Nature dictated to a heathen poet, and by which she guided his predecessors, is the actual rule of God in the testimony borne by Omnipotence to Truth—

"Nec deus intersit nisi dignus vindice nodus."

A needless miracle would be unworthy of Heaven, and incredible to enlightened men. But further, in the working of the miracle itself, all that can be done, by human power and ordinary means, is commanded to be done. If water is to be made wine, the water-pots are filled by the hands of men. It would have been as easy to create the wine at once, but, in that case, the law of nothing superfluous and nothing defective would have been violated. If the leper is to be cured, he must wash seven times in the waters of Jordan and be clean. If the withered hand is to be restored, the patient must himself make an effort to stretch it forth. If dead Lazarus is to be raised, men must roll the stone from the mouth of the sepulchre, and when he comes forth bound hand and foot, they must unloose the grave clothes and let him go. Even in the two miracles of feeding the multitudes, the baskets of fragments

I am aware that there are some seeming exceptions to the rule which I wish to demonstrate. For instance, the rain that would cheer the thirsty ground in a season of drought, and save the fruits of the earth for the use of man, may return from the clouds to the ocean, or fall upon the sterile sand; while at other times the labors of the husbandman may be deluged from on high, and his wealth swept away by the torrent that gathers among the hills. But that the rain is wasted even on the sea or the sand, would be far too much for us to affirm; and no believer in Providence will find difficulty in rightly interpreting the variations and hazards that attend the cultivation of the soil. In fact, we know too little of meteorology to decide what advantage may arise to the whole globe from the phenomena of the sky; but we know enough of nature's works to be assured that every phenomenon must accord with the law for which I am now contending.

Up to this point, however, I have said nothing of

were not superfluous but intentional; because, besides being gathered up for future use, they afforded the Saviour the very opportunity he sought and planned, of inculcating care and frugality that nothing might be lost. This lesson was, in fact, a Divine proclamation of the law in question—that in God's doings there is no deficiency and no redundancy. The same Being, who created food in the lonely place for his hearers, could have done so for himself when he hungered in the wilderness; but the miracle was not needed, and, therefore, it was not performed. Thus the multiplying of the loaves and fishes, which, at first glance, appears to contradict our principle, really goes to confirm it in every particular.

the proportion between the powers and the work of intellectual and moral beings. This, indeed, is the very question which we are required to determine. But the condition of man is obviously excluded from our argument: for it is upon man's condition as a conclusion that the entire argument is intended to bear; and except in so far as we can perceive, in the present state of humanity, indications of primeval perfection, the whole of our inductive evidence must necessarily be analogical.

Excluding man, then, the nearest approach to intelligence in terrestrial nature is the instinct of animals; and it was once my purpose to relieve the tedium of our present investigation by adducing illustrations of the law now under discussion, from that interesting field of Natural History. Such a course, however, would prolong this address beyond all due bounds; and I, therefore, content myself with a general statement of fact—that, while every instinct that is necessary for the comfort and preservation of brutes is bestowed upon them, they possess, in their natural condition, none that are superfluous. When any of them are domesticated by man, they are rendered so far artificial, and some of their original instincts may thus become useless. But in their wild state these instincts are indispensable. The dog, which now turns round several times before lying down to sleep, is only practicing in domesticity the gyration by which his ancestors hollowed out their lairs in the wilderness. Hence we have the strong analogy of instinct to add to the evidence already adduced, that the powers of every creature are exactly proportioned to the work which that creature has to perform.

Whether or not there are any spiritual beings between man and God is a question which mere philosophy does not enable us to decide. It is the general belief of the human race that there are such beings; and the testimony of Scripture, which reveals the existence of Angels as a matter of fact, coincides with this general belief. A dogma of revelation, however, cannot be used as a link in any chain of purely philosophical argument. All the aid that we are entitled to claim from the Bible is the fact, that everything therein declared concerning the nature of Angels is in perfect accordance with the conclusions which I have already drawn from the attributes of God. These pure spirits are always represented as busy in the service of their Lord. They rest not day and night. Their devotion to God is entire; and not a single hint is dropped to the effect that any of their power is ever kept back, or diverted, from the work that their Creator has assigned them. Their duty and their delight is to employ all their faculties, at all times, and in all their available strength, in the service of

Him from whom these faculties were derived. There is not one circumstance in this representation which conflicts with our notions of justice and propriety. Everything is exactly as we should judge it to be from the relation of spirits to the Father of Spirits. We feel assured that, if there really are spiritual creatures superior to ourselves, the law of their duty to God is precisely that which the Bible describes.

But though it would be illogical to pass, in an argument of this kind, from reason to revelation when reason fails us, it is manifestly lawful to rest upon well-attested historical facts, whether these are facts of Jewish or of Gentile history. Now, the appearance and ministry of angels I hold to be historical facts. No candid critic can confound them with the fables of Greek and Roman mythology. I am, therefore, justified in assuming that the general belief of mankind on the question of superhuman spirits is correct; and this being the case, my rational apprehension of Ethical relations assures me that their power and their duty are most scrupulously proportioned to each other, and that such of them as have not abused their spiritual liberty do fulfill their duty to the very letter.

If, then, the law of exact correspondence and proportion between power and work extends over both the highest and the lowest of God's creatures, it were most unreasonable to imagine that man, who stands between the brute and the angel—a compound of the animal and the spiritual—can be exempted from the rule that applies to the animal and the spiritual alike. When we find man doing other work than his God's, the rational inference is—not that his capacities of intelligence, feeling and will are insufficient for the attainment of the end of his being—but that his original condition has undergone a change—that he has abused his moral freedom, and is a rebel against the law. Most unwarrantable it were to suppose that the law has been abrogated in his favor, or even in the smallest degree relaxed. The existence of the law is manifest; its foundation lies in the relation of creature to Creator, from which relation the creature, man, can claim no exemption; and what can be more reasonable than the employment of God's own gifts in God's own service, and in nothing else? The law, indeed, is not only reasonable but supremely benevolent; for the only solid happiness lies in strict obedience to its commands. Bird, beast, reptile, fish and insect are all happy in the exercise of their instincts, and the use of their powers. To do the bidding of the Most High constitutes the blessedness of angels. And every human being, who has abandoned his rebellion and returned to his allegiance, is forward to proclaim that he never knew substantial enjoyment till now.*

We thus find that the conclusion for which I have been contending is supported both by considerations *a priori*, and by examples drawn from every region of nature; and we are abundantly warranted in affirming that every creature has a work to do for his Creator, and that his Creator has furnished him with powers precisely proportioned and adapted to that work. The work does not exceed the powers, and the power is not greater than the work. It is thus manifestly impossible that a creature can ever do more than his duty to God; and consequently, in

* The aphorism of the Great Teacher is at once natural and true: "Unto whomsoever much is given, of him shall much be required." The whole law of the creature lies in this Divine announcement; and it is fully illustrated in the parables of the talents and pounds, and of the wise and foolish virgins. I cite another passage of Scripture to the same effect. It is this: "Whether ye eat or drink, or whatsoever ye do, do all to the glory of God." The precept is quite general. All that we do, even to the most common and necessary actions, is to be done—not unto ourselves, nor to any other creature—but to the glory of Him who made us. I need not, therefore, dwell on that aspect of the command. I prefer to inquire what is meant by eating and drinking to God's glory. Manifestly, not that which some suppose—namely, eating and drinking with thankful hearts. Doubtless, gratitude to Providence for our daily bread is a good thing, and a duty; but I cannot believe that that idea exhausts the significance of an expression so remarkable as the one in question. The real and full meaning of the passage is clearly this: that, in the matter of eating and drinking, we are bound to eat and drink of such things, and in such quantities, and at such times, as will maintain our powers, both mental and bodily, in the highest possible state of efficiency and endurance, for the service of our Lord and Master. Excess, on the one hand, and abstinence, on the other, are equally derelictions of duty, unless some absolute necessity, or some higher duty, intervenes to modify our practice.

case of failure, he can make absolutely no compensation—in case of arrearage, future payment is utterly beyond his own ability. Any *accumulation* of creature merit is an obvious absurdity; and so there never can be a surplus to atone for a single moment's idleness, or a single moment's relaxation beyond that rest which may be really required by the creature's constitution. If the faculties of an angel are nobler than those of a man, the angel has a more arduous task to perform; and both man and angel are bound, by the relation of creature to Creator, to employ continually their whole available power in their Creator's service. And more than this, they are also bound to take good care that no abuse of any sort—neither of improper exertion, nor of sensual indulgence—shall diminish their power, on the one hand, and that no exercise shall be neglected, on the other, which may increase its efficiency, according to the appointed law of such increase. There is no allowance for indolence, or carelessness, or irregular activity; far less for positive perversion. Should the duty of any man ever call for over-exertion, and consequent destruction of power, the sacrifice is required by his present abnormal condition. In a perfect state no such demand can be made; for no such sacrifice can be necessary.

Now, keeping fully in mind what I have thus

demonstrated respecting the duty of man, and the relation of his original power to that duty, let me attempt the proof of a second proposition, which, after its establishment, I shall ask you to connect with the first, so as to draw a conclusion from a comparison of the two.

My second proposition is the following: The scheme of God in creation and providence is progressive: not in the sense of proceeding by trial and error, as the schemes of men do; but in the sense of proceeding from perfection to further perfection.

If you reflect on the connection of cause and effect, as exhibited in the universe, you will find that no cause is followed by one effect only. There may, indeed, be, and there usually is, one effect of which the given cause is more particularly the antecedent; but, in addition to this prominent effect, there are also minor and collateral effects which must be ascribed to the same cause; and each of these effects becomes, in its turn, a cause destined to produce so many separate series of new effects, and so on *ad infinitum*. The propagation of effects is thus like the propagation of a race of animals or vegetables from a parent stock. In fact, the indefinite propagation of organized creatures is just an instance of that causal progression whereof I am now speaking. Perpetual progress is, therefore, a necessary result of the great law of cause and effect. From the Almighty First Cause, as from the centre of power, streams of causation are forever radiating, and forever widening, in a multiplied efficiency, towards the outer regions of unlimited space, and through the endless ages of infinite time.

But I will not rest the demonstration of our second proposition, any more than I did that of our first, upon mere *a priori* considerations. It will be more interesting to you, and quite as much to our present purpose, if I can lead you, by a brief induction of particular cases, to a satisfactory establishment of the general law.

From an examination of the rocks which compose the crust of our earth, and the organic remains that are therein imbedded, we find that this world has undergone a succession of wonderful changes, in which creation after creation, each perfect in its kind, has been destroyed, and by which the globe has been gradually prepared for the comfortable habitation of the human race. The geological history of the earth is one of the sublimest retrospects that scientific research affords. Through the mighty and mysterious ages of the past, mortal and irrational creatures have been employed as the precursors and pioneers of the rational and immortal, and we believe that, after one change more, all of the latter that has become liable to death shall be re-endowed with immortality, and not a bone of man shall continue in the dust. The ground shall give up its human dead—not in fragments and fossils for the instruction of superior beings—but living, and to live forever, in their renovated abode.

Again—rising from the earth to the heavens, we discover there appearances which go to prove that there is a similar creative progression in other worlds besides our own. There seems sufficient truth in the nebular hypothesis, to warrant the conclusion that the realms of space contain systems in all stages of formation, from the most chaotic and rudimentary, up to those which we are wont to call perfect. Creation nowhere springs at once to the highest beauty, but unfolds its glories by degrees. The eternal Maker lays his commands on matter, and He, to whom a thousand years are as one day, guides it obediently, through countless ages, to its destined end.

Returning to our own earth, we there perceive the same law of progress in detail, which we have already observed in the general. The life force in animals and vegetables builds up bodies for them by a gradual process of assimilation and growth, and matures in them the germs of future generations; so that from one tree may spring a forest, and from one pair the population of a planet; and in the higher region of human personality, the intellectual, æsthetical, and moral powers work out, by continued effort, the advancement of science, and art, and liberty. In spite of reverses and vicissitudes, and transmigrations from one country to another, civilization, and knowledge, and government are perpetually moving forward, on the whole. "Antiquitas seculi, juventas mundi." The human family—as a family—are not only older, but wiser, and better, and happier *now* than ever they were since first they peopled the earth. The comparative barbarism, that has overrun some ancient fields of refinement, is more than compensated by the higher culture of others, and by the gladness of many a primeval wilderness that has been made to rejoice and blossom as the rose.

On the whole, then, and not to weary you by further induction, I venture to affirm that progress from one degree of perfection to another is a law which the Almighty has been pleased to enact for His own operations, and for the continued felicity of His rational and responsible creatures. Of the progression that is visible in material things, I need say nothing more. The notice of it was necessary only to fortify our second proposition, as I did the first, by analogy. I crave your particular attention to the progress of the intellectual and moral universe, and to the fact, that the very nature and necessities of spiritual creatures compel us to believe, that the same increase of power and enjoyment which we discern in the human race is also a law to every rational and accountable subject of the King of kings. We know enough of our own souls to feel assured that, were it not for the perishable bodies wherein they dwell, their capacity of improvement is indefinite; and consequently, in the case of pure spirits, there can be no limit to the accumulating strength acquired by perpetual exercise.

Let us now connect our two propositions, and see what conclusion will result from them.

Every creature has a work to perform, and power enough, but not more than enough, to perform it. As the power increases by continued exercise, the work increases in exact proportion. The ratio of the two is always a ratio of equality. We have thus, in the universe of God, a perpetually augmenting power and a perpetually augmenting work—a continued progress which will never have an end—a vast procession of intelligence and virtue, ever mounting and ever hastening towards loftier heights of knowledge, righteousness, and holiness. Should any creature, or any company, in that universal march, stop short in the exercise of their faculties—sit down indolently by the way, or absolutely commence to struggle backward against the advancing host—thus wasting their strength in vain perversity, tell me, I pray you, what consequences would follow? The grand procession hurries on with ever-growing power and speed. The loiterers and mutineers are left behind, losing vigor, both of intellect and will, every moment of their stay. The distance between them and their former fellows is ever, ever widening, while their own capacities for good are ever, ever diminishing. Their perdition is deepening by a double acceleration. The case is clearly a hopeless one—hopeless, most hopeless—unless God himself can open up an avenue of hope!

Turn aside with me, therefore, and gaze on this great sight—this wondrous procession of angel and archangel, cherubim and seraphim! Onward and upward tread incessantly these unfallen sons of God! Failing in no duty since they first were made, they have ever been mounting from glory to glory, and from strength to strength. Their intelligence has been perpetually expanding, and their knowledge has been perpetually augmenting. Their affections have been continually deepening, and fresh objects of affection have been continually supplied as their capacities enlarged. Their moral sense has been always acquiring new vigor, their will has been always growing more resolute, and their lapse into disobedience has been evermore becoming less and less possible.

It is a marvelous panorama that we have now before us! Not the sons of Genius, struggling upwards, with panting breath and many a slip, to some mythic immortality on the heights of Olympus or of Helicon—not these, but the sons of Almighty God, bright with eternal youth, and strong with ever-growing strength—tasked to the full, but never overstrained—exultant in their ascent as the eagle in its flight—marching up the highway to the heaven of heavens, while the splendors of the holy place cast on their path a brighter glory than the sunshine, and the chorus of triumph swells from rear to van of the magnificent procession! Not the stars of heaven—not suns, with their planetary trains, sweeping onward through space—not galaxies rushing in cycles that baffle computation, yet still returning whence they came as the appointed ages roll away—not these grand orbs, but spirits immortal, each more precious than a thousand stars, advancing forever and forever towards that Sanctuary where sitteth the Father of Spirits—"high-throned above all height"—unapproachable, yet "altogether lovely," and still disclosing new beauties to His children as they rise!

But where is man in this majestic progress—what place holds he in the universal host? He, too, was destined to a post in the procession, and, though the last of God's children, was not the least in His regard. Angels would not have disdained his company, nor would his voice of joy have been discordant with their song. He, no less than they, would have proceeded from perfection to perfection—his capacity, like theirs, forever growing and forever full! But man is confessedly a deserter from the army of the Lord of Hosts. The most orthodox believer bears no stronger testimony to this fact, than does the zealous reformer, who frequently would compensate for the scantiness of his creed by the extent of his philanthropy. Forsaking the ranks of Heaven, and in league with the rebellious, man has met the fate of the dupe in his apostacy. He now strays and struggles in the wilderness—struggles with its entanglements, seeks a home in its spots of transient verdure, and strays further and further from the way of life. His faculties have been perverted, his affections have been misplaced, and his will has been depraved. Sloth has enervated him, passion has wasted his vigor, and he either sits down or retrogrades, while the universe hastens on! The interval between where he is and where he ought to be is perpetually lengthening, and the cumulative power, that was due to his continued exercise in holiness, is irrevocably gone. Never, even though he desired it, can he overtake his former companions; neither, though he could overtake them, has he now the strength to keep pace with them in their accelerating march. They are now stronger than they were, and he, to all true good, is weaker. Desolate and helpless as Israel in the House of Bondage—desolate and helpless as the captives who hung their harps upon the willows, and wept by the rivers of Babylon—desolate and helpless as the daughter of Zion bowing in sorrow beneath the palm-tree—desolate and helpless as the prodigal who, far from love and home, would fain have fed on unclean husks—desolate and helpless as these, he sits him down—and who shall bear him across the space that intervenes between him and the post he should have held—who shall replace the strength that he has squandered in iniquity, and supply the power that he ought to have gained in the practice of righteousness? Manifestly, not himself; for at no period had he more power than he needed, and now he has far less. Manifestly, not an angel, nor an army of angels, for, though they may pity the apostate, they have no power to spare. Manifestly, no created thing—manifestly, none but the Omnipotent—none but One who is absolute and independent—One who can interpose, with the fullness of underived and unclaimed might, to seek and to save the ruined.

"How charming is divine Philosophy!" How

charming at all times, but especially how charming when she thus leads us to the portals of Divine Revelation, and the response of the Holy Oracle harmonizes with the voice of Reason! There really is an Almighty Redeemer—an Omnipotent One that lays hold on wretched man, and bears him to where he should have been in the universal march; and who, from the riches of His grace, can furnish more than all the energy that man has lost! "Who is this that cometh from Edom; with dyed garments from Bozrah? this that is glorious in his apparel, travelling in the greatness of his strength?" It is I—"I that speak in righteousness, mighty to save."

From the hour that this Deliverer espoused our cause, the door of hope was opened, and the free favor of Heaven, descending to bless this blighted earth, prevented its degenerating into a pandemonium. Divine mercy, that heretofore might have been heard of by the hearing of the ear, but which no eye had yet seen in actual operation—this new attribute of Godhead, like a new system in immensity, was disclosed to the admiration of angels and men. From that blessed hour, captive after captive began to be released. Death, the avenger, was made the herald of eternal life, and the grave of the now mortal body become the gate of glory to the still immortal soul. From that hour, the noble work of emancipation—emancipation to light and power as well as freedom—has been going on; multitudes of the rescued have been welcomed to the celestial throng; and we believe—for our natural expectation is unquenchable, and the oracles of prophecy assure us—that a day of triumphant restoration is drawing nigh. It is written! it is sealed in heaven! and the fullness of time shall reveal it all! And when the great day shall come at last, there shall be such a merry-making in the universe as has not been since of old the morning stars sang together; for the crowning act of a new and nobler creation shall have been brought to a close. The Celestial Host, whose glory lighted the plains of Bethlehem, and whose anthem echoed along its hills, shall again unfold their splendors and take up their song; and Earth below, no longer mute as in the beginning, but vocal throughout all her realms, shall send back her joyous response to the gates of Paradise. The mountains shall break forth into singing; the fields shall clap hands on every side; the glorious strain shall ring in the harping of the woods; streams shall murmur praise as they flow; and ocean shall uplift his music of many waters in concert with the quiring winds; the stars shall peal notes of gratulation from their spheres; the great sun shall roll through all his deep tones of rejoicing; the ransomed themselves shall lead the mighty jubilee with "blest voices uttering joy;" while the vaulted sky, like a high temple-roof, shall resound the glad chorus of a renovated world, and a race at length made free!

www.ingramcontent.com/pod-product-compliance
Lightning Source LLC
LaVergne TN
LVHW011208110826
845150LV00006B/1370

* 9 7 8 1 4 2 5 5 1 8 2 6 4 *